HOW TO BUILD A 'CRETAN SAIL' WIND-PUMP

for use in low-speed wind conditions

by
R.D. Mann, A.I. Agr.E., M.L. Biol.,
Agricultural Officer
Gambia Christian Council
West Africa

INTERMEDIATE TECHNOLOGY PUBLICATIONS

The printing of this publication has been made possible by generous grants from three donors who wish to remain anonymous.

Practical Action Publishing Ltd
27a Albert Street, Rugby, CV21 2SG, Warwickshire, UK
www.practicalactionpublishing.org

First published 1979
Reprinted 1983
Reprinted March 1986
Reprinted February 1992

Reprinted by Practical Action Publishing
Rugby, Warwickshire UK

ISBN 9780903031660

A catalogue record for this book is available from the British Library.

Since 1974, Practical Action Publishing has published and disseminated books and information in support of international development work throughout the world. Practical Action Publishing is a trading name of Practical Action Publishing Ltd (Company Reg. No. 1159018), the wholly owned publishing company of Practical Action. Practical Action Publishing trades only in support of its parent charity objectives and any profits are covenanted back to Practical Action (Charity Reg. No. 247257, Group VAT Registration No. 880 9924 76).

CONTENTS

List of plates *Page*

List of illustrations and tables

PART 1

Introduction

At the end of 1974, the Gambia Christian Council started a small-scale village-level agricultural programme. The rains in the Gambia occur from mid-July to mid-October, and during the nine-month dry season there can be no field cropping without irrigation. The drought conditions from 1968 to 1977, when average rainfall decreased from 51.6 inches to 35.8 inches per year, caused a shortage in domestic food supply, and any method of producing food or cash crops in the dry season is of considerable value to the rural community.

The G.C.C. agricultural programme has consisted of introducing vegetable production on a planned basis, involving the use of livestock-proof fencing and the sinking of 4' diameter concrete-lined wells at the rate of two wells per acre. The wells range in depth from 15' to 35', and all water extraction is by bucket and rope, the water being carried by hand to the vegetable plots. Each village project is planned for an area of one to two acres under vegetables each season, one acre being sufficient for 25 families.

At their current stage of development, this labour-intensive method of irrigation by hand is not a limiting factor to the success of these projects as the availability of water from October to May, together with improved methods of crop husbandry, are the most important inputs.

However, when one considers future development possibilities, and in particular the present need for tree-planting in all Sahelian and near-Sahelian countries, a mechanical means of water lifting could have far-reaching results in making such schemes feasible, and this has been the background thinking which prompted the development of the Gambia Christian Council Wind-Pump.

Meteorological Considerations

With reference to the wind frequencies given for three stations in The Gambia in Tables 1, 3 and 4, it will be seen that there is no wind for 27% to 36% of the time, wind speeds of over 12 miles per hour occur for only 3% to 9% of the year, and the balance of 61% to 64% of the time has wind speeds up to 12 miles per hour.

So for any practical use to be made of wind power, a windmill design is required which can start and operate in low wind speeds varying from 5 to 10 miles per hour.

Hydrological considerations

The river Gambia is tidal for a distance of 150 miles or more from the coast, and therefore water for irrigation can only be taken from the upper reaches of the river. The ground water level varies from 15' to 80', the deeper wells being mainly in the eastern end of the country.

Most dry-season vegetable growing areas are low-lying with water tables varying from 15' to 25', but their location would require a high windmill tower of 50' to 60' in order to reach the available wind. However, there are some areas, suitable for dry-season cropping, with a water table of no more than 20' depth at the end of the dry season, and it was decided to build a windmill to operate a simple piston lift-pump.

Wind-pump design

The design was based on the information given in the ITDG publication, ***Food from Windmills,*** and from practical advice given by Mr Peter Fraenkel, ITDG Power Project Engineer, in September 1976. The design of the 'Omo' windmill, developed by the American Presbyterian Mission in Ethiopia, was studied in detail. The 'Omo' windmill worked mainly in a wind regime of speeds from 8 to 15 miles per hour, and it had problems associated with turntable rotation and directional stability of the wind-wheel into the wind. Since our winds are much lighter than those of the Ethiopian situation, it was decided to make the wind-wheel 16' in diameter and with six arms to permit the use of two, three, four or six sails as required.

With reference to drawings A and B (given in Part 2) the drive shaft turns in three bearings, two of oil-impregnated hardwood, the rear bearing being a self-aligning ball-bearing which also takes the axial thrust. The adjustable crank permits pump strokes of 5⅞", 7" and 8⅛" to be used. The pump is the same as used on the 'Omo' windmills; it is a 3" diameter piston operating in a 16' p.v.c. cylinder, and it is connected to the crankshaft via a universal joint. The turntable is fitted with four sealed roller-bearing units, which run on a ½" wide, 12" diameter bearing track.

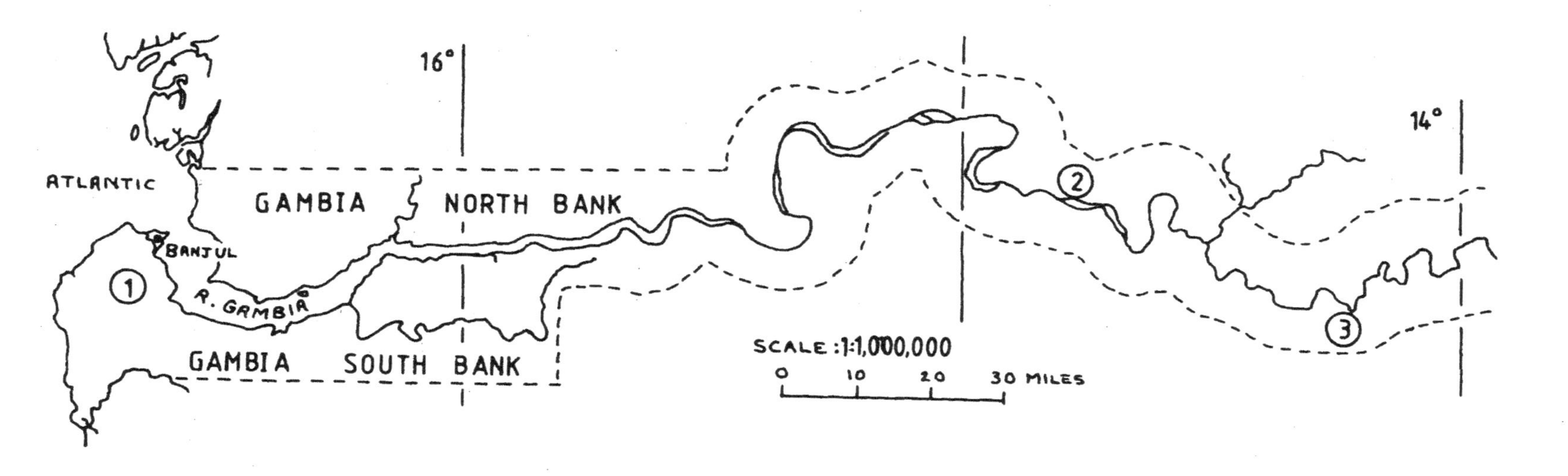

Table 1 Wind frequencies for site 1 at Yundum

WIND SPEED	J	F	M	A	M	J	J	A	S	O	N	D	Average
Calm	22	17	13	10	12	20	30	33	41	48	48	28	22%
1-12 miles per hour	70	71	71	72	73	66	61	59	55	50	50	66	64%
12-24 miles per hour	8	12	16	18	15	14	9	8	4	2	2	6	9%
Over 24 miles per hour	0	0	0	0	0	0	0	0	0	0	0	0	0

Table 2 G.C.C. Wind-pump output on four selected days

Date:	Wind:	Test period	Less stops:	Net period	Height lifted:	Wind-run:	Gallons pumped
4.4.78	NE-NW	0830-1500	5 mins	6 hrs 25 mins	13' 4"	33 miles	1162 galls
5.4.78	NE-NW	0830-1804	67 mins	8 hrs 27 mins	13' 4"	49 miles	1937 galls
6.4.78	NW-NE	0830-1800	11 mins	9 hrs 19 mins	13' 4"	62 miles	3861 galls
7.4.78	NW	0830-1800	5 mins	9 hrs 25 mins	13' 4"	64 miles	3342 galls

Table 3 Wind frequencies for site 2 at Georgetown

WIND SPEED	J	F	M	A	M	J	J	A	S	O	N	D	Average
Calm	35	34	35	32	24	20	27	33	38	46	50	42	35%
1-12 miles per hour	58	61	60	65	71	75	70	64	60	53	49	53	61%
12-24 miles per hour	7	5	5	3	5	5	3	3	2	1	1	5	4%

Table 4 Wind frequencies for site 3 at Basse

WIND SPEED	J	F	M	A	M	J	J	A	S	O	N	D	Average
Calm	–	39	29	25	18	15	28	33	50	47	55	55	36%
1-12 miles per hour	–	55	66	73	78	80	70	64	49	52	43	41	61%
12-24 miles per hour	–	6	5	2	4	5	2	3	1	1	2	4	3%

The tail-fin has been built high in relation to its width to obtain maximum leverage in the wind-stream. The tail unit is of articulated design, with control ropes operated below the wheel to put the wheel into or out of the wind. The position of the tail-fin can be adjusted on the tail boom to obtain balance across the turntable bearings and so provide the best response to changes in wind direction.

The tower is three-cornered and made in three sections for ease of transport and erection. The tower feet have plates welded on to the bottom of the legs, and are sunk into the ground to a depth of 4'. When erecting the tower, a spirit level is held across the bearing track at the top, the position of the legs being adjusted in turn until the track is precisely level in all directions, and the holes are then filled in with tightly packed soil. The height of the tower is 23' from ground level to the bearing track.

The 'main' sails are made of heavy-duty marine canvas, and the 'starter' sails are lightweight cotton. The inner and outer sail corners are fitted with rubber loops (details of which are given in the construction data in Part 2), and it takes about six minutes to either fit or remove the sails.

The wind-wheel is provided with struts and perimeter-wire tension adjusters so that the wheel can be made quite taut, and thus any tendency of the wheel arms to flex during gusting winds is avoided.

As will be seen in the construction drawings in Part 2, the various windmill components can all be unbolted to allow for easy maintenance and modifications. After completing the workshop construction, all the mild steel parts are coated with a marine anti-corrosive paint to give long-term rust protection.

The construction of this prototype machine was carried out on a part-time basis, as and when other field duties permitted, in the Ministry of Agriculture engineering workshop at Yundum Experimental Station. The construction commenced in March 1977 and was completed by the end of September 1977.

Field testing

To facilitate testing, the wind-pump was sited adjacent to a 7½' deep concrete water reservoir tank at Abuko, 2½ miles from Yundum. The top of the tank was at ground level, and the total lift could be kept constant, the water lifted being measured in drums and then drained back into the reservoir.

There were a few scattered trees of up to 35′ in height, but none within 100 yards of the wind-pump, and at 150 yards and beyond the tree-line was mainly oil-palm. The obstruction to wind-flow caused by this vegetation was considered to be fairly typical of that which would be found at other suitable pumping sites. Initial field trials were carried out in November and December 1977.

The wheel was first fitted with six large sails, all the same size, with the outer sail corners (corner C shown in drawing U) held by tight rubber loops to the outer sail hooks on the perimeter wires. With this arrangement there was no directional stability in wind speeds of 8 m.p.h. and above. The sail pattern was changed to three large sails, corners fitted as before, and this provided better wind-wheel stability and less sail-flapping.

During these first trials, the arms of the wind-wheel were not fitted with support struts. It was soon found that the wind-wheel was too flexible in gusting winds, and struts were then fitted to each arm which solved this mechanical problem and made the wheel completely rigid.

The tail boom design was also changed. The initial tail boom was made of 1″ x 1″ angle iron, 59″ total length, but this allowed twisting of the boom and caused the tail-fin to shake under wind pressure. A new tubular tail boom was designed (as shown in drawing Q), and to counter-balance the extra weight of the struts on the wheel it was then possible to fit the tail-fin a further 7″ away from the turntable.

The result of these alterations to the wheel and tail, together with the use of three large sails, was improved sensitivity to changes in wind direction, but there was still some sail flapping and the wheel would tend to over-run at wind speeds above 9 m.p.h. During sudden high-speed gusts, the back of the sails would occasionally hit the tower legs, and there was still the tendency for the wheel to move to the right-hand side around the tower as the speed of the wind and the wheel increased.

The large sails provided adequate sail area to drive the wheel, but during light winds the starting performance was poor. To provide more starting torque, three small-size sails were designed with a no-load 'angle of attack' to the wind of approximately 19°.

During uniform wind conditions, in February and March 1978, the sail arrangements of 3 'main' sails, 2 'main' sails and 2 'starter' sails, and 3 'main' sails with 3 'starter' sails were compared, and it was found that for a wind speed range of 5 to 10 m.p.h. the best overall performance was obtained with 3 'main' sails and 3 'starter' sails.

After further trials, it was found that sail flapping and any tendency of the main sail to hit the tower legs, could be completely eliminated by tying the leading edges of the sails tightly against the wheel arms, and a further considerable improvement in wheel rotation and pumping output was obtained by fitting lengths of rubber from the preceding wheel arms to the outer sail corners, full details of which are given in Part 2 under the section entitled 'Sails'. This provision of an elasticated connection on each sail trailing edge also had the effect of governing the wheel speed during high gusts, and thus prevented the wheel from over-running.

The final improvement involved counteracting the tendency of the wheel to swing to the right-hand side around the tower as the wheel-rotation speed increased. To do this, the boom catch (see drawing P) was released, and in the first instance the tail boom was tied in a position 10° to the left (looking forwards to the wheel). The tendency was now for the wheel to move progressively around the tower to the left-hand side; a final position of about 7½° to the left was then found to give the best directional stability of the wheel, and the boom was fixed in this position for the remainder of the trials.

For the pumping tests, two drums (each of 48.4 gallons capacity) with screw-type outlets at the bottom were placed on the reservoir wall. The pump inlet was connected via a 1½″ diameter p.v.c. pipe to a 1¼″ factory-made brass foot-valve placed in the water tank. The pump outlet was taken by a 1½″ diameter p.v.c. pipe to the drums. Wind-run was measured by an integrating cup anemometer mounted on a pole at a height of 6½′.

With reference to Graph 1, the wind-speed was calculated from the time taken to fill a drum and the wind-run as shown by the anemometer at the commencement and completion of each drum-filling period. On some of the test days, the total lift (measured from the water reservoir surface to the pump outlet, vertically) was 13′ 4″, and on others it was 14′ 1″, so all output figures have been equated to foot-gallons for ease of comparison. The work done in foot-gallons per hour was calculated by noting the time required to fill each drum. As one drum was filled to the top, the

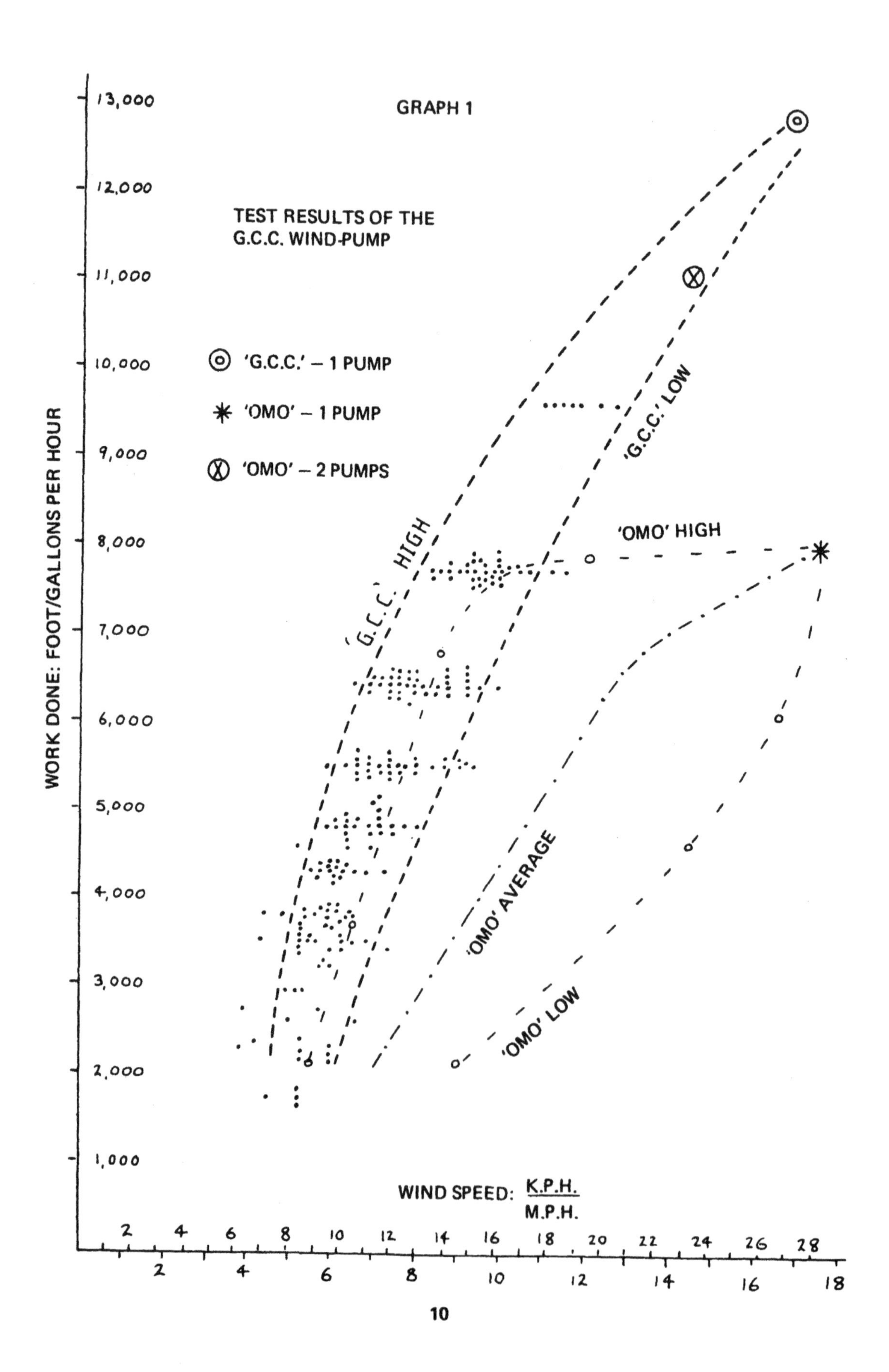
GRAPH 1
TEST RESULTS OF THE
G.C.C. WIND-PUMP
'G.C.C.' – 1 PUMP
'OMO' – 1 PUMP
'OMO' – 2 PUMPS
WORK DONE: FOOT/GALLONS PER HOUR
13,000
12,000
11,000
10,000
9,000
8,000
7,000
6,000
5,000
4,000
3,000
2,000
1,000
'G.C.C.' HIGH
'G.C.C.' LOW
'OMO' HIGH
'OMO' AVERAGE
'OMO' LOW
WIND SPEED: K.P.H. / M.P.H.
2 4 6 8 10 12 14 16 18 20 22 24 26 28
2 4 6 8 10 12 14 16 18

outlet pipe was transferred to the other drum, the first drum being drained back into the tank, so that continuous recording could be carried out. Since it was intended to obtain a realistic picture of the pumping capability, the recording was not interrupted if the wind-wheel stopped turning for short periods during variable light winds, and all meaningful recorded figures have been included in the graphs.

Again with reference to Graph 1, a total of 196 separate drum-filling measurements have been plotted, together with 19 various-quantity output readings taken specifically to indicate performance at low wind speeds. The pump stroke used during these tests was 8¹/₈". The sail arrangement was changed for brief periods during the trials, but the majority of readings were taken using three 'main' sails and three 'starter' sails.

The pumping output of the 'Omo' windmills, as given in the publication ***Food from Windmills,*** has been put on the same graph for ease of reference, and it will be seen that the highest recording of the G.C.C. wind-pump (with a single pump) is greater than the output of the 'Omo' windmill fitted with two pumps.

The wind speed required to start the G.C.C. wind-pump from rest was calculated to be between 5.2 and 5.6 m.p.h., and once started the wind-wheel continued to run in a steady wind down to 4.5 m.p.h. The usual wind-speed range at Abuko was found to be 0 to 10 m.p.h., and the highest recorded wind-speed during the trials was 16.9 m.p.h.

The output obtained on four selected days is shown in Table 2, the 'net period' and 'wind-run' figures being inclusive of those periods when no pumping occurred due to insufficient wind.

Conclusions

1. The design arrived at in the G.C.C. wind-pump has produced a good pumping output under specific low-lift conditions at sea-level.
2. There is no visible adverse wear in the turn-table bearings or on the bearing track after a period of 10 months.
3. Very little wear has occurred in the oil-impregnated shaft and big-end wood-block bearings, which appear to be both durable and suited to local construction.
4. The universal-joint design appears to be mechanically sound; providing it is lubricated properly it should last for a while.
5. The safety straps and rollers which were subjected to buffeting during initial sail trials, show no signs of distortion.
6. The track guides function well in facilitating turntable rotation, but in future models they should be located on the other side of the bearings to slide round the inner track ring, so that wear will occur only on the easily replaceable inner ring.
7. Four bearings for the turntable are not required, three being quite sufficient, placed 120° apart, and the bearing load will then be equalised. This modification should be made on all future machines.
8. The 1³/₈" diameter drive shaft has shown no sign of fracture or bending, but it would be better to use a 1½" diameter shaft for this size of wind-wheel.
9. The marine canvas and light-weight cotton used for the sails have not shown serious wear so far, but these materials will not last very long and the sails should be made of a yacht-sail material such as 'Dacron' to provide a good life under constant use.
10. The articulating tail boom design is not essential when the sails can be easily fitted and removed each day, and it is considered better for economy if this complex mechanism is omitted. Future designs should allow the boom angle to be adjusted while tests are carried out to obtain the best directional stability, following which the boom can be bolted in position and a rope left hanging down from the boom to pull the wind-wheel out of the wind when removing the sails.
11. The total cost to build this prototype wind-pump amounted to D2,400-00 for materials. (Note: D4-00 = £1.00.) On a local production basis, the cost of labour might be about D600-00, which would give a total of D3000-00 per machine. This figure compares favourably with the price of approximately D10,000-00 to import a comparable wind-mill from the U.K.
12. The G.C.C. wind-pump has so far lifted water a height of only 14 feet. The maximum height that water can be raised using a lift-pump is not much more than 21 feet, and this limits the use of a lift-pump to riverside pumping and those few places with ground water near the surface.
13. In view of the point made in 12, above, further testing of the G.C.C. wind-pump will include trials with a force-pump to see if it can lift water to a height of 45 feet, and if this is possible it may have a wider application for village level development.

ACKNOWLEDGEMENTS

The construction of this machine would not have been possible without the considerable assistance given by Mr P. Cham, Agricultural Engineer, Mr A. Senghor, Workshop Foreman, Mr A. Hughes, Technician, and Mr S. Peter, Welder Technician, at the Agricultural Engineering Workshop, Yundum.

St. Peter's Catholic Mission at Lamin kindly allowed us to use their workshop facilities for machine drilling of components, and their Engineer, Mr R. Houston assisted with modification work.

Valuable welding advice was also provided by Mr C. Marsh, VSO Engineer attached to the Vocational Training Centre in Banjul.

Grateful thanks are also due to Mr Dumbuya, Farm Manager at Yundum Experimental Station, who provided labour on several occasions during erection and subsequent modification of the wind-pump.

We are grateful to Mr Mbye-Njie for permission to use the water reservoir, located on his farm at Abuko, for the pumping trials.

Acknowledgement is made of the considerable initial development work carried out by staff of the American Presbyterian Mission at Omo Station in Ethiopia, and the practical advice given by Mr P. Fraenkel, ITDG Power Project Engineer.

Particular mention must be made of Mr P. Hutchinson, United Nations W.M.O. Agro-Climatologist in the Gambia, who gave encouragement throughout all the development stages, and who, with staff of the government meteorological department, supervised and evaluated the pumping trials in the field.

We also remember, with appreciation, Rev. C. Awotwi Pratt, Chairman of the Methodist Mission, who gave approval for expenditure of Mission funds at the commencement of the project, and are also grateful for subsequent contributions towards the project costs by the Ministry of Economic Planning in Banjul and the Commonwealth Secretariat in London.

R.D. Mann, A.L. Agr. E., M.I. Biol.
Agricultural Officer,
Gambia Christian Council.

References

Food from Windmills by Peter Fraenkel, ITDG, November 1975.

The Windmill Experiment by Mr P. Hutchinson, Hydromet, Banjul, June 1978.

PART TWO — CONSTRUCTION DETAILS

INDEX

Notes

i. In the attached drawings, adequate construction detail is provided of all the wind-pump components, but it must be noted that certain items are omitted from particular drawing views, where clarity of the described part is required.

ii. All angle iron referred to in the construction description is of ¼" web thickness except where otherwise stated.

A.1. Turntable Frame

Refer to drawings C,D,E, and O.

1. Cut 2 lengths of 1½" x 1½" angle, item 1, 36¾" long. Drill ½" diameter (9/16") holes, items 2a, at one end of each frame side piece, remembering that one is left-hand and the other right-hand, and also drill the bearing-box bolt-holes 3/8" diameter (7/16"), item 3.

   At a point 9½" from the other ends of item 1 cut 'V' notches of 47°, bend upwards, and weld each at 45° angle.
2. Cut 2 pieces of 1½" x 1½" angle, item 4, 10¾" long, and drill the bearing-box bolt-holes 3/8" diameter (7/16"), item 5, in each.
3. Weld items 4 to items 1 to form the frame bed.
4. Cut 2 pieces of 1½" x 1½" angle, item 6, 4" long, as legs for the rear bearing support.
5. Cut 2 pieces of 1½" x 1½" angle, item 7, 6" long, as legs for the middle bearing support.
6. Cut 1 piece of 1½" x 1½" angle, item 8, 10¾" long, and 2 pieces of 1½" x 1½" angle, items 9 and 10, 11¼" long, these pieces to be slightly oversize in length to provide a jam-fit assembly to facilitate welding.

   In the rear bearing support, item 10, drill holes of diameter and spacing to suit the rear bearing selected. ***Note: for this prototype wind-pump, a self-aligning ball-bearing complete with grease-packed housing was used.*** In the downwards leg of the angle, item 10, three holes of ½" diameter, item 11, must also be drilled for the tail boom control ropes to pass through. In the middle and front bearing supports, items 9 and 8, drill ½" diameter (9/16") holes, item 12, 4" apart.
7. Weld the rear legs, item 6, in place; then jam-fit item 10 between the rear legs with upper surface of the angle at 2¼" from upper edge of main frame sides and inclined upwards/forwards at 5° as shown in drawing C. Weld in position.

   Note: a 5° inclination is obtained by a 1" vertical rise over a length of 12".
8. Weld the middle legs, item 7, to the main frame sides. Jam-fit item 9 between the middle legs, also inclined at 5° upwards/forwards in line with the upper surface of the rear bearing support, item 10, and weld in place.

   Similarly, jam-fit item 8 between the foremost part of the frame side pieces, at 5° inclined upwards/forwards with its upper surface in line with the rear and middle bearing supports, and weld in position. ***Note: the drive shaft is inclined upwards at 5° to ensure that the wind-***

Note: The item numbers referred to throughout the following text are shown as circled numbers on the drawings attached.

wheel and sails have adequate clearance from the tower.

A.2. Turntable Bearings

Refer to drawings, F, J, and K.

1. Cut a piece of 2" x ¼" flat mild steel, item 13, 6" long, and drill bolt holes, item 14, 3/8" diameter.
2. Cut a piece of 2½" x 1½" box section, item 15, cut away both sides at one end by ½" depth as shown, and drill a 5/8" diameter hole through the box to take the bearing bolt, item 16.
3. Weld the box, item 15, to the flat plate, item 13.
4. Cut a piece of 1¼" x ¼" flat mild steel, item 17, 2¼" long and drill a 5/8" diameter hole in it for the bearing bolt. ***Note: this piece acts as a guide to prevent the bearings moving sideways off the turntable bearing track.***
5. For this prototype machine, a sealed bearing, item 18, was used, of 2" outer diameter, 9/16" wide, with 13/16" inner bore diameter.

   The bearing was fitted on to the 5/8" diameter bearing bolt, item 16, on a pipe sleeve, item 19, with distance piece, item 20, made of pipe, and distance washers, items 21 and 22, either side of the track guide, item 17.

Notes:

(a) the actual thickness and position selected for the washers and distance piece must be such as to locate all four turntable bearings precisely on the ½" wide turntable bearing track within an accuracy of plus or minus 1/16" sideways.

(b) the turntable safety straps and rollers, which bolt on beside each turntable bearing, are discussed in the tower head section below as they are related to the safety ring fitted in the tower.

A.3. Wind-wheel Drive Shaft

Refer to drawings C and G.

1. The crankshaft is made in 3 sections, the front shaft, item 23, the crank, item 24, and the rear shaft, item 25. All three parts are made of ¾" inner diameter galvanised steel pipe of such outer diameter that it can be lightly force-fitted, using a heavy wooden mallet, inside 1" inner diameter galvanised steel pipe, the result being a thick-walled hollow shaft and crank journal of 13/8" outer diameter.
2. Cut the front shaft, item 23, to a length of 17¾".
3. Cut the rear shaft, item 25, to a length of 12¼".
4. Cut the crank journal, item 24, to a length of 5".
5. Cut 2 pieces of 2" x ¼" flat mild steel, item 26, 7" long; drill the two ½" diameter holes accurately; mark and punch the centre of each piece and scribe a circle round the centre of 1½" diameter to facilitate alignment during welding.
6. Make up a jig out of angle iron to support both the front shaft, item 23, and the rear shaft, item 25; bolt the crank-bolt plates, item 26, together, with the scribed circles facing outwards; draw the front and rear shafts together to butt against both sides of the bolted plates, and having aligned the shaft ends within the scribed circles, weld the shafts to the crank-bolt plates.
7. Cut 2 pieces of 2" x ¼" flat mild steel, item 27, 9" long; drill the six ½" diameter holes, item 28, accurately in each piece; mark and punch the centres where the crank journal is to be welded and scribe a circle 1½" diameter around each centre point.
8. A big-end bearing of 1½" width and 1½" diameter was selected for this machine, and to ensure its correct location on the journal, 4 flat washers, item 30, and a phosphur-bronze bush, item 31, were fitted on to the journal with 1/8" diameter split pins provided behind the outer washers.
9. Using the same jig as referred to in 6 above, bolt the crank-bolt plates with shafts, item 26/25 and item 26/23, to the crank arms, item 27; position the crank journal item 24, complete with fittings, accurately within the scribed circles (a large G-clamp is useful to clamp the pieces together) and weld the crank journal to the crank arms.

Notes:

(a) alignment of the shafts is not easy, but the jig will help together with careful welding, and the alignment between front and rear bearings should be accurate to within 1/16".

(b) the front shaft flange is discussed in the section on the wind-wheel as it is related to the wheel hub.

Underneath view of turntable frame showing bearing boxes.

A view of the turntable.

Turntable in operation on tower.

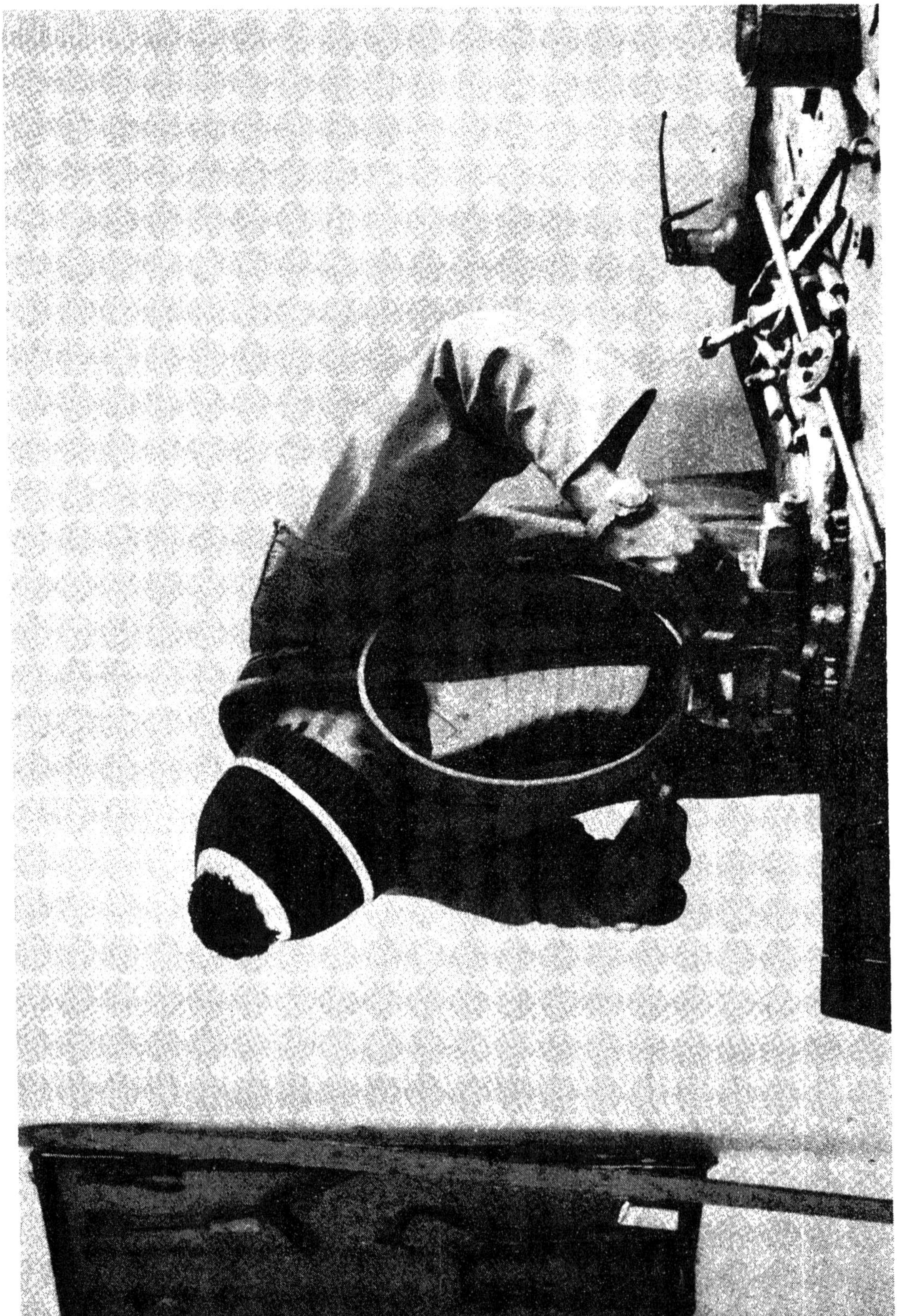

Making a turntable bearing track.

A.4. Crankshaft Bearings

With reference to drawings C and H, the front and middle crankshaft bearings are made as follows:

1. Obtain a hardwood, remembering that the greater the density the better the bearing life (bearings on this machine were made of 'Mampato', a West African tree harder than 'African Mahogany').
2. For each bearing, cut 2 blocks, item 32, each 2" x 3" x 6" but oversize by $^1/_{16}$" in 3" to allow for shrinkage.
3. Clamp the blocks together in pairs; drill the shaft hole, item 33, 1$^7/_{16}$" diameter (to allow for shrinkage), and holes $^9/_{16}$" diameter for the securing bolts, item 34; punch mark the blocks in pairs on adjacent faces before unclamping.
4. Half fill a 5 gallon drum with used engine oil. Place the wooden blocks in the oil with a weight on top to keep them submerged. Heat the oil steadily, observing the moisture release from the blocks indicated by the air bubbles rising to the oil surface. When the bubbles have reduced in size to that of a pin-head, remove the fire, and leave the oil and blocks to cool overnight. Remove the blocks after 24 hours when they will have absorbed all the oil they can hold.
5. Make the bearing bottom plate, item 35, of 6" x 3" x ¼" mild steel; drill bolt holes $^9/_{16}$" diameter. The bearing top plate, item 36, is of 6" x 3" x $^1/_8$" mild steel, drilled with two $^9/_{16}$" bolt holes and a central hole for fitting the screw-type grease-cup, item 37. Drill a grease hole $^1/_8$" diameter in the upper bearing block leading to the bearing surface.
6. The grease-cup can be welded on to the top plate, item 36, or threaded into the upper bearing block. When fitting the wood-block bearings on to the shaft, use shims between the blocks for initial running-in rather than trying to enlarge the bearing surface if it is found to be tight.

Note:** **The rear shaft bearing, item 38, is a factory-made self-aligning ball-bearing, complete with grease-packed housing, and it is levelled with the shaft by inserting distance pieces and/or shims between the housing and support. This bearing is of 1½" internal diameter and is locked on to the 1$^3/_8$" shaft by a tapered screw-sleeve, which then takes the axial thrust from the wind-wheel.

B.1. Tower Head

Refer to drawings I, J and K.

1. Cut 3 pieces of 1½" x 1½" angle, item 39, 24" long with 30° angles at each end, and weld together to form an equilateral triangle.
2. Make two circles from 1" x ¼" flat mild steel, one of 12" inner diameter, item 40, the other of 12½" inner diameter, item 41, so they will fit one inside the other to provide a ½" wide bearing track of 13" outer diameter. When joining the ends of the 1" x ¼" mild steel, they must be cut to make 'V' notches on both sides, welded accurately, and then filed smooth on all surfaces.
3. Place the track, item 40/41, centrally on top of the triangular frame, and take measurements for cutting the 3 track support struts, item 42, from 2" x ¼" flat mild steel, such that these struts will project inwards beyond the inner edge of the track by not more than $^3/_8$" (to ensure that safety straps do not touch the struts as the turntable rotates). Weld the struts, item 42, in place with their upper surfaces level with the upper surface of the triangular frame.
4. Cut 9 pieces of 1¼" x 1" x ¼" mild steel, item 43, and drill a hole of $^3/_8$" diameter centrally in each. Place these pieces symetrically around the track with 3 of them central on the support struts and the other 6 equidistant at 60° apart as shown in drawing I. Weld all 9 pieces, item 43, to the outside face of the outer track ring, item 41.
5. Punch mark centrally through each of the $^3/_8$" diameter holes in the track legs, item 43, on to the triangular frame and the support struts; remove the track and drill each point to $^7/_{16}$" diameter.
6. Weld $^3/_8$" diameter studs, item 44, into each of the track legs, item 43, and the track can now be bolted down on the frame.
7. Make another circle from 1" x ¼" flat mild steel, of 12" inner diameter, for the safety ring, item 45.
8. Cut 3 safety ring support legs, item 46, of 1¾" x 1¼" x ¼", and drill each with a $^7/_{16}$" diameter hole.
9. Cut 6 distance pieces of 1" x 1¼" x $^3/_8$", item 47, and 3 distance pieces of 1¾" x 1¼" x $^1/_8$", item 48; drill $^3/_8$" diameter holes in the latter.
10. Turn the head-frame upside down, bolt the legs, item 46, and the distance pieces, item 48, on the support strut studs; place the

distance pieces, item 47, symetrically on the head-frame sides, and accurately position the safety ring, item 45, in line vertically with the inner track ring, item 40. Weld the safety ring on to the 3 legs and 6 distance pieces.

11. The safety straps with rollers, item 49 are made and fitted as follows: Assemble the turntable and bearings, and place this unit on the turntable track. Make 4 roller units, each consisting of a 1½" length of ¾" outer diameter galvanised pipe placed on a ½" diameter bolt, the end of the latter being welded on the 1½" x 1¾" x ¼" roller plate as illustrated in drawing K. Make 4 safety-strap legs of 1" x 3/16" flat mild steel, drilled to fit on the bearing-box bolt, bent at 90°, ensuring that the legs are sufficiently far out not to catch on the head frame as the turntable rotates. Bolt each safety leg on one side of each bearing box; clamp the roller unit on to the lower end of the leg, leaving 1/8" clearance between the roller and safety ring, and weld together.

B.2. Tower Upper Section

Refer to drawings A and L.

1. Cut 3 pieces of 1½" x 1½" angle, item 50, 78" long, for the tower upper-section legs.
2. Cut 6 pieces of 1½" x 1½" angle, item 51, 24" long, with 30° angles at each end, and weld up 2 triangular frames; one frame is used as the tower upper-section, 'bottom frame', the other as the 'top frame' of the tower middle section. Clamp these two frames together face to face, and drill 6 holes, 9/16" diameter at the spacing shown, for the attachment bolts.
3. Cut 3 pieces of 1" x 1" angle for lateral reinforcement, item 52.
4. Place the legs, item 50, at right angles inside the corners of the head frame, item 39, and the 'bottom frame', item 51, and weld all round. Weld the lateral members, item 52, inside the tower legs. On each of the 3 sides of the tower, ¼" diameter reinforcement rod, item 53, is welded on to the outside of the frames and the lateral members to provide diagonal strength.

B.3. Tower Middle Section

Refer to drawings L,M and N.

1. Cut 3 lengths of 1½" x 1½" angle, item 54, 120½" long, for the middle-section legs, and 3 lengths of 1½" x 1½" angle, item 55, 130½" long, for the lower-section tower legs. Drill 9/16" diameter holes on alternative sides of each leg for the ½' diameter attachment bolts (i.e. 2 bolts per leg) at the spacing shown, so that the middle-section legs overlap outside the lower-section legs when bolted together.
2. Place 2 middle-section legs, item 54, inside 2 corners of the 'top frame' (already constructed in B.2. 2. above) with the legs 55¼" apart measured between inside corners of the angle at the lower ends. Support the 'top frame' in a position of 7° away from perpendicular to the common plane of the two legs (to obtain the right inclination of the tower legs), then weld the legs to the frame.
3. Cut 3 lateral members, items 56a, 56b and 56c, of 1" x 1" angle, of the length required to fit in the positions indicated, and weld them inside the legs.
4. Complete the other two sides of the tower by welding the 3rd tower leg inside the 'top frame', with the other two lateral members, item 56c, in position to obtain the correct inclination of the tower leg, followed by welding the other lateral members, items 56a and 56b inside the angle legs.
5. Diagonal strength is obtained by welding ¼" diameter reinforcing rod, item 53, to the outside of the 'top frame' legs, and lateral members, on all three sides, in the positions shown.

B.4. Tower Lower Section

Refer to drawings M and N.

1. Lay the middle-section of the tower on its side, with blocks to hold it off the ground throughout its length by 2".
2. On to the middle-section legs, bolt 2 of the lower-section legs, item 55.
3. Fit a piece of 1" x 1" angle for the bottom lateral member, item 57a, between and inside the legs so that the outer width measurement at this point is 76".

   Note: the legs of the lower-section are splayed outwards by 2" each at the bottom to provide additional tower stability, and the bottom lateral must be fitted first.

4. Weld lateral member, item 57a, in position first, then fit and weld lateral members, items 57b and 57c.

Making upper tower section frame.

Fitting bearing track on head frame.

5. Bolt the 3rd lower-section leg on to the middle-section. Cut 2 more bottom lateral members, item 57a, and weld in position, followed by the remaining lateral members, items 57b and 57c.
6. Unbolt the lower-section from the middle section; weld on to the legs and the lateral members ¼" diameter reinforcement rod, item 53, for diagonal bracing, in the positions shown, on all 3 sides.

   Note: the upper ends of the diagonal bracing must be welded inside the legs.
7. Weld a piece of 4" x 4" angle, 12" long, item 58, on to the bottom of each lower-section leg. This will allow the wind-pump tower to be erected in compacted soil without any need for concrete tower footings.

C. Tail Unit

Refer to drawings O,P,Q and R.

Note: The actual method of fabricating the tail unit is not critical, but the construction detail for this unit is given in full, since from this description many alternative designs will become apparent. The important points are:

i. to make the tail boom last as long as possible, without causing imbalance of the turntable unit, to obtain maximum leverage,
ii. to make the tail of lightweight material to allow the use of a long boom,
iii. to make the tail fin high in relation to its width, so that it confronts a greater cross-section of the wind-power for leverage, and,
iv. to provide some means of varying the horizontal angle between the tail fin and the wind-wheel to allow for experiment during trials to obtain the best directional stability.

With reference to the drawings provided, the tail unit on this machine is designed with a pivoting boom, operated by control ropes which lead down within the tower frame, for pulling the wind-wheel into and out of the wind. It is constructed as follows:

1. Make up a rectangular pivot-frame, item 59, of 1½" x 1½" and 1" x 1" angle, with attachment legs, item 60, of 1½" x 1½" angle, and brace pieces, item 61, 1" x ¼" flat mild steel. The braces, item 61, are attached with 3/8" diameter bolts at their upper ends on to the pivot-frame, and their lower ends are held by the left- and right-hand bearing bolts. The legs, item 60, are attached with ½" diameter bolts, item 2b, through the bolt holes, item 2a, at the back of the turntable frame, item 1.
2. Two pairs of ½" diameter pivot holes are drilled in the top and bottom members of the pivot-frame, one pair on the centre line and the other pair 3" offset to the right. ***Note: only the pair of offset holes were used for the boom pivot in these trials.*** Additional pivot-pin bearing-support surface is provided by welding drilled support pieces, item 62, of ¼" thick mild steel, inside the upper and lower angle members. Before assembly, a 3/8" diameter hole is drilled in the right-hand leg to take the boom stop-arm bolt, item 63; and the right-hand end of the lower angle is cut away as shown at 64a. The pivot-frame and legs are then welded together.
3. The boom stop-arm, item 64b, is made of 1" x 1" angle, 15" long. It carries the 'starting' pulley and its bracket, item 65a, the boom-stop, item 66a, and spring-loaded boom-catch, item 66b. The 'stop' pulley, item 65b, is bolted to the left-hand vertical member of the pivot-frame. The pulley attachment bolts are ¼" diameter. The tail control-ropes roller, item 66c, is fabricated of ½" galvanised pipe with washer-flanges brazed on each end, running on a 3/8" diameter bolt, and with legs, of ¾" or 1/8" mild steel attached on the turntable rear-bearing bolts. A rope guard, item 66e of ¼" diameter steel rod is welded across the roller arms to keep the ropes in place.
4. The holes, item 11, through the rear crank-shaft bearing-support, item 10, have pieces of rubber pipe inserted into them to prevent undue wear on the control ropes. Rubber, 1/8" thick, is wrapped around the contact surface of the boom-stop, item 66a, to provide a cushioned effect for the catch mechanism.
5. The tail control ropes, item 66d, are made of 3/16" diameter nylon. The boom-catch rope is attached at the end of the catch. Both the 'start' and 'stop' ropes are tied onto the boom-frame vertical member, item 67a, just below the catch position. A strong strip of vehicle-tyre inner tube, of 2" x 1/8" section, is wrapped several times around the left-hand ends of the upper and lower angle members of the pivot-frame, item 59, to act as a stop for the boom-frame when the wind-wheel is put out of the wind.
6. The tail-boom, item 68a, is made of 1" outer diameter galvanised pipe, 69½" long. To

facilitate the attachment of the tail fin and boom-frame, thin-wall (1/32") square-section mild-steel box pieces, items 68b and 68c, of 1" x 1" internal measurement, are coated inside with anti-rust paint, then push-fitted over each end of the boom pipe, and brazed in place.

7. The foot of the boom-frame vertical member, item 67a, of 1" x 1" angle, is attached by a 3/8" diameter, item 67b, of 1" x 1" angle, has a bracket, item 67c, also of 1" x 1" angle, welded at the rear end, to take the ¼" diameter tension bolt, item 67d, for the 1/8" diameter galvanised brace-wire, item 67e. A hole is drilled into the top of the boom, 11½" from the end, to take the ¼" diameter anchor-hook, item 67f, for the brace-wire.
8. The horizontal member, item 67b, and the front end of the boom, item 68c, are drilled in line to take the ½" diameter pivot-pin, item 67g, the latter being provided with washers and split-pins top and bottom. The horizontal member, item 67b, and the vertical member, item 67a, are joined by welding, and a drilled support-piece, item 67h, is welded on top of the horizontal member (with the pivot-pin in place for alignment) to give added bearing surface against the pivot-pin at this point. Washers are also placed on the pivot-pin between the pivot-frame and the boom-frame.
9. A ½" diameter coil-spring, item 69a, of suitable strength, is held at its front end by the leg, item 69b, attached to the right-hand pivot-frame bolt, item 2b, and at the other end to the leg, item 69c, bolted to the boom. This spring provides an over-centre action to assist the boom in its travel towards the boom-catch when the 'start' rope is pulled.
10. Tail fin position-adjustment holes of 3/8" diameter, at 2" spacing, are drilled in the boom end, item 68b. The tail fin, item 70a, made of 3/16" thick plywood, 96" tall by 18" wide is attached by 3/32" diameter bolts within the frame, item 70b, of ½" x ½" x 1/16" mild steel angle, the latter being brazed together. Cross-pieces, item 70c, each made of 2 pieces of ½" x ½" x 1/16" angle, are brazed inside the other (after first coating internal surfaces with anti-rust paint) for added strength, and each cross-piece is brazed to the side of the frame.
11. Four pieces of ½" inner diameter lightweight pipe, item 70d, each ½" long, are brazed one to each corner of the frame; these pipe rings facilitate attachment of the 1/16" diameter galvanised bracing wires, and have rubber inserts to prevent wear on the wire at these points. Metal hooks, item 70e, of ¼" diameter reinforcing rod, are brazed to the outer ends of the cross-pieces, to take the bracing wires, item 70f, fitted from top and bottom of the tail fin, and also the bracing wires, item 70g, fitted across and diagonally between the cross-pieces.

    To obtain sufficient tension on the bracing wires, item 70f, simple tension adjusters, item 70h, are made with a 3/16" diameter eye-bolt and a foot-piece of ½" x 1/8" section mild steel as illustrated.
12. To give extra rigidity to the connection between the tail fin and the boom, support pieces, item 70i, of 1" x ¼" flat mild steel, are placed both sides of the tail fin and bolted together through the plywood with ¼" diameter bolts; the two support pieces on the right-hand side (those illustrated in the side view in drawing Q) are 8" long, and those on the other side are 23" long.

    The boom is connected to the tail fin assembly by two 3/8" diameter bolts, item 70j, which pass through the centre points of the supports pieces, item 70i.

D. The Wind-wheel

Refer to drawings S, T and U.

1. Cut 2 pieces of ¼" thick mild steel plate with 6 sides, 8" across flats; drill 12 bolt holes of 7/16" diameter at the spacing indicated, clamping the plates together to ensure alignment, and a ¾" diameter hole through the centre of both plates. One of these plates is the shaft flange, item 71a, and the other plate is the wheel hub, item 72a.
2. A ¾" length of ¾" outer diameter galvanised pipe, item 71b, is pushed into the centre hole, and welded flush with the rear surface, of the shaft flange, item 71a; this acts as a location stub when fitting the wind-wheel in place. The shaft flange, item 71a, is then welded centrally on the end of the front shaft, item 23.
3. Shaft flange reinforcement is provided by 6 triangular webs, item 71c, of 1½" x 2" x ¼" mild steel; a 1½" long piece of mild steel pipe is first cut lengthwise into two halves, placed together on the front shaft behind the

Back view of the turntable showing, on the far left, the bolt holes to which the legs of the tail unit are attached.

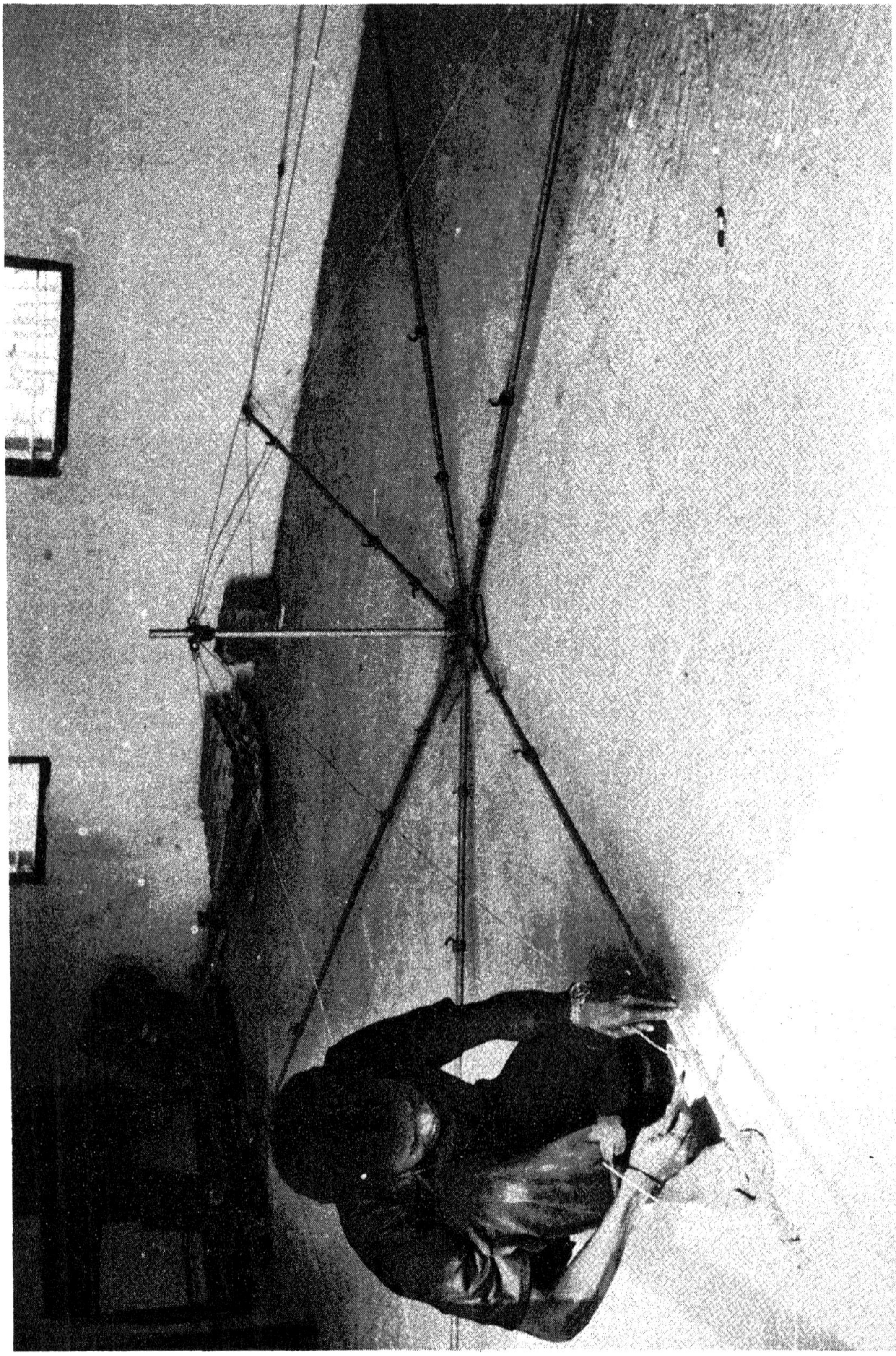

Assembly of wind wheel.

Fitting the wind wheel.

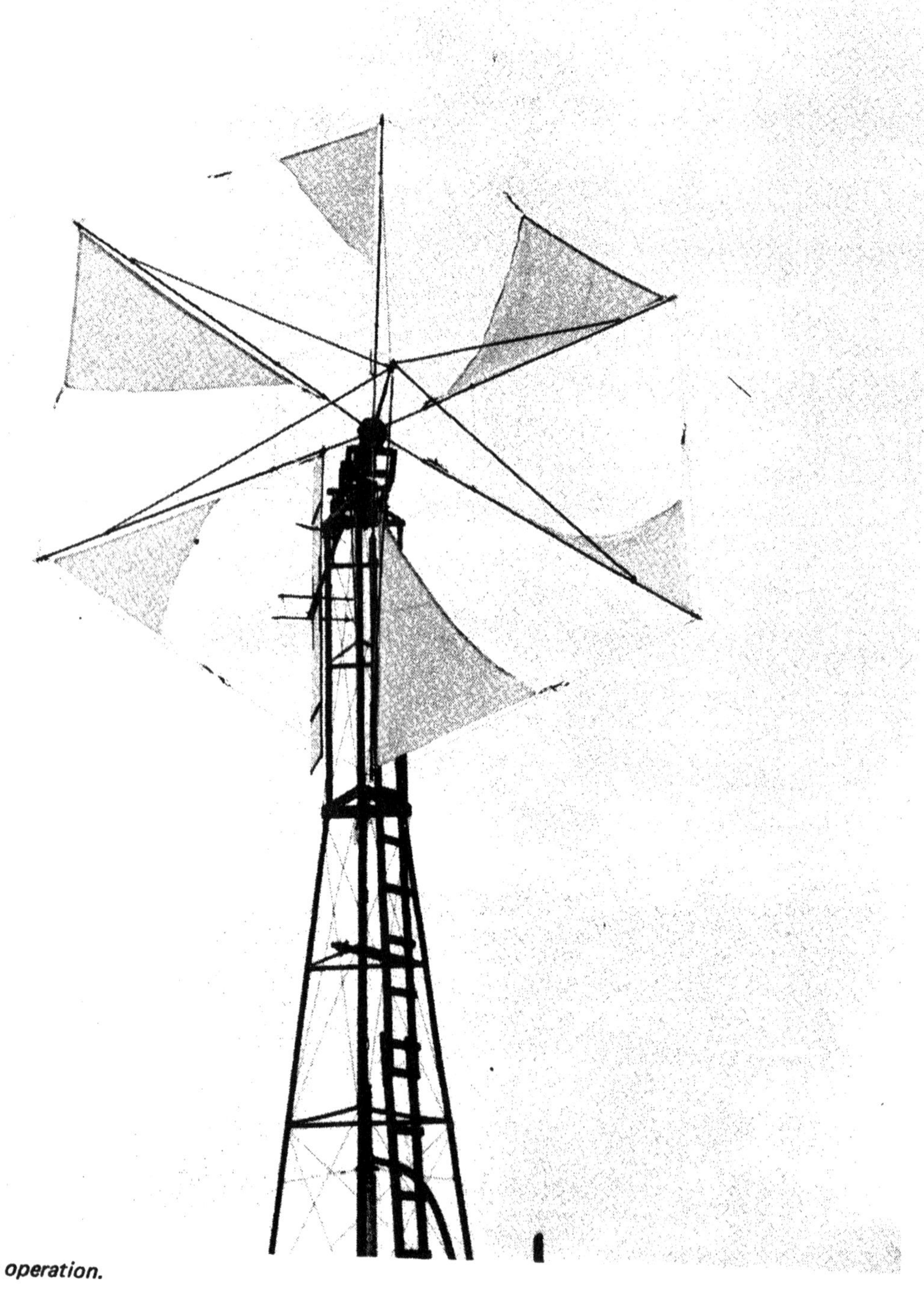

The sails in operation.

flange and welded tightly along the lengthwise joints and at the front edge to the flange rear face; the 6 webs are then welded to the pipe sleeve and to the flange, with light tack welds to reduce distortion.

4. Weld a 1" length of 1" inner diameter galvanised pipe, item 72b, centrally on the front of the wheel hub, item 72a.
5. The wheel centre-shaft, item 72c, is made of galvanised pipe; an 18" length of ½" inner diameter pipe is force fitted a distance of 7" inside a 22" length of ¾" inner diameter pipe, the end of the latter pipe being inserted into the socket, item 72b, and both these joints welded.
6. The strut adjuster, item 72d, made of ¾" inner diameter pipe, 2½" long, is made to slide fit on the front end of the wheel centre-shaft, and is secured by a ¼" diameter split-pin through one of the 5 adjustment holes provided in the shaft at ¾" spacing. Six eyes of ¼" diameter reinforcing rod are welded around the strut adjuster.
7. The 6 wheel arms, item 72e, and the 6 wheel struts, item 72f, are made of 5/8" inner diameter thin-wall galvanised conduit pipe. Each wheel arm, item 72e, is flattened at the inner end over a distance of 4" to 5", bent at this point by 3° forward, and drilled with two 3/8" diameter holes for the attachment bolts, item 72g.
8. The outer end of each arm is threaded, and carries a ½" length of pipe coupling, on to which a loop of ¼" diameter reinforcing rod is welded to form an anchor, item 72h, for the wheel perimeter-wire, item 72i. Each arm also carries 4 hooks, 2 for sail attachment, one for brace-wire attachment, and the other to take the wheel strut. All hooks are of the same design, being made of ¼" diameter steel rod, welded on to a 1 1/8" length of thick-wall mild steel pipe of such diameter to be a slide fit on the wheel arm. Each hook unit is fastened by a 3/32" diameter steel pin through the hook pipe and wheel arm. The sail corner and inner hooks project sideways from the arm, while the strut and brace-wire hooks project forward from the arm.

   Note: to facilitate bolting the struts to the arms, the strut hooks should be in the form of an eye.

9. Between the corner sail-hook, item 72j, and the perimeter-wire anchor, item 72h, a ¼" diameter hole is drilled sideways through each arm to take a ¼" diameter steel adjuster-hook, item 72k, complete with lock nuts. The end of each perimeter-wire is attached to its own adjuster-hook on the preceding arm.
10. A wheel strut, item 72f, is provided on each of the 6 wheel arms. Each end of the strut is slotted and drilled to take 3/16" diameter bolts to secure it to the strut hook on the wheel arm and to its corresponding eye on the strut adjuster.
11. The 6 wheel perimeter-wires, item 72i, are made of 1/8" diameter galvanised mild steel wire. The outer sail hooks, item 72l, are made of ¼" diameter reinforcing rod with an eye at one end through which the perimeter-wire passes in the form of a loop to hold the hook securely. The outer sail hooks are positioned 56" from the anchor loop to the bend of the hook, and the hook shank is bound to the perimeter-wire with 1/16" diameter galvanised wire to hold it rigid.
12. When the perimeter-wires have all been fitted to their anchor points and to the adjuster hooks, the adjuster nuts are progressively tightened in turn around the wheel until all slack in the strut connections is taken up and the perimeter-wires are quite taut. After tightening, the wheel arms should be approximately 3° forward of the plane of the wheel hub. The brace-wire, item 72m, also of 1/8" diameter galvanised mild steel wire, is then wound around each brace-wire hook, item 72n, from one arm to the next, tightened by hand, to prevent sideways movement of the arms.

E. Sails

Refer to drawings B and U.

1. The 'main' sails, item 73, are made from heavy-duty marine canvas, with a 1" wide hem, in which 3/16" diameter nylon rope is placed, on all three sides. These ropes are tied tightly through the edge of the canvas at each corner, so that the sail cloth cannot slide back on the ropes under wind pressure. If the cloth corners are not secured to the ropes in this way, the sail shape will alter under wind-load, thus reducing the efficiency of the sails and the pumping output will be reduced.
2. Corner A of the main sail has a short loop of rope to go on the corner sail hook. Corner B has a tight loop of rubber (lorry type inner

tube) of ¾" x 1/8" section to fit on the inner sail hook. Two rope ties, of 1/8" diameter cotton rope, passing through the sail canvas hem behind the hem rope, tie the leading edge close to the wheel arm. These latter ties are very important for correct sail operation.

3. During initial field trials, the outer sail corner C was tied to the outer sail hook, item 72l, on the perimeter-wire by a tight rubber loop, but this arrangement caused the wheel to over-run accompanied by sail-flapping during gusting wind conditions. A long rubber loop of ¾" x 1/8" section was then tied from the outer sail corner C to the perimeter-wire anchor on the preceding arm, and this arrangement provided progressive governing of the wheel speed as the wind speed increased and eliminated flapping of the sails.

   Note: the side dimensions of the main sails are such that in the no-load position, i.e. with the sail corners fitted by insufficient wind to turn the wheel, the 'angle of attack' at mid-point of the trailing edge is approximately 7½°.

4. The 'starter' sails, item 74, are made of light-weight cotton, locally known as 'greyball' and used for sails on fishing boats. As will be seen from the dimensions of the 3 sides, the starter sails provide a greater 'angle of attack' to the wind of about 19° at the mid-point of the trailing edge at no-load. These sails, fitted on alternative arms around the wheel, are provided specifically to assist in starting at low wind speed, and they also help to maintain wheel rotation between gusts in light wind conditions.
5. The starter sails also have a 1" wide hem, through which a 3/16" diameter nylon rope passes on all 3 sides, all rope ends being tied tightly through the material at the sail corners as on the main sails.
6. The hem rope is brought out in a short loop at corner A of the starter sail to go on the corner sail hook. The rope running inside the sail edge A to B extends to a tight rubber loop of ¾" x 1/8" section to fit on the inner sail hook. The rope running inside the sail edge A to C extends 15" alongside the perimeter-wire and is then tied to a simple strand of ½" to 1/8" section rubber with its end looped over the perimeter-wire anchor on the preceding arm. A 1/8" diameter cotton-rope tie is passed through the sail material behind the hem rope at corner B to keep the leading edge of the starter sail against the wheel arm.

Notes:

(a) a short length of rubber tube is pushed on to the inner and corner sail hooks to avoid rapid wear of the sail attachment loops.

(b) in steady winds of 6.8 m.p.h. (11 k.p.h.) and above, the starter sails need not be used, but they are required where winds are variable and light.

F. Pump and Connecting Rod

Refer to drawings M, V, W, X, and H.

1. The pump, item 75a, is a 'Dempster' 3" inner diameter p.v.c. cylinder pump with two leather piston washers. The bucket and valve components are of gun-metal, and the top and bottom of the cylinder are threaded for 1¼" diameter inlet and outlet. The cylinder is 16" in length.
2. The pump inlet, item 75b, is 1¼" diameter galvanised steel pipe, proceeding via a gentle 90° pipe bend (not shown) to a 1¼"/1½" socket into which is threaded a 1½" diameter thick-walled p.v.c. suction pipe. The end of the suction pipe is fitted with a 1½"/1¼" socket to take a factory-made brass foot-valve.
3. The outlet, item 75c, is via a standard pipe 'T' fitting, above which a further pipe extension, item 75d, is fitted to take a 'Dempster' brass stuffing-box, item 75e.

   The outlet pipe is also 1¼" diameter galvanised pipe, and to facilitate the wind-pump testing was connected via a 1¼"/1½" socket to a length of 1½" thick-walled p.v.c. pipe to reach the drums used for measuring the output.

4. The pump foot mounting consists of a socket, item 75f, with legs welded each side and fixed with ½" diameter bolts inside two pieces of 1¼" x 1¼" galvanised steel angle, item 75g, the latter being held by four ¼" diameter 'J' hook bolts across the lateral members, item 56c.
5. The pump top is held centrally by a fabricated clamp-ring, item 75h, made of ¾" x 1/8" section mild steel, with 3 legs, item 75i, each of which takes a ¼" diameter 'J' hook bolt, item 75j. Each hook bolt passes through a tension-adjuster leg, item 75k, which in turn is connected by ¼" diameter reinforcing rod to one of the tower legs, item 54.
6. The pump is held down in the socket, item 75f, by means of two 1/8" diameter galvanised wires, item 75l, attached to ¼" diameter 'J'

hook bolts, item 75m, which pass through the angle supports, item 75g.

7. The pump connecting rod, item 76a, is fabricated of 5/8" outer diameter conduit pipe, as illustrated in drawing W, and joined to the pump rod, item 76b, by a connecting link, item 76c, made up of 2 pieces of 1" x ¼" mild steel and two nuts.
8. The connecting rod is kept in alignment by a cross-head, item 77a, of 2" x 1" timber, held by 2 plates, item 77b, of 1" x ¼" mild steel, and four ¼" diameter 'J' hook bolts, item 77c, across the lateral members, item 56a. A 1" diameter hole is drilled centrally through the cross-head, item 77a, and an easily replaceable guide block, item 77d, provided with a 5/8" diameter clearance-hole, is screwed on top of the cross-head. Both pieces of timber are hardwood.
9. The adjustable crank journal, item 24, can be bolted in 3 different positions to give a piston stroke of 5 7/8", 7", and 8 1/8" respectively, and the length of the connecting rod, item 76a, is calculated accordingly.
10. A universal joint, item 78, shown in detail in drawing X, joins the connecting rod, item 76a, to the crank rod, item 79a. The replaceable wearing parts of the universal joint are the loose washers at top and bottom of the pipe guide, and a bush, made of conduit pipe, which fits inside the pipe sleeve bearing.
11. The crank rod, item 79a, is also made of 5/8" outer diameter conduit pipe, connected top and bottom by pipe couplings which are tack-welded at one end, and fitted with 3/32" diameter retaining pins at the other end to facilitate disconnection when necessary.
12. The construction of the big-end bearing is shown in drawing H. The blocks, item 80a, of this split wood-block bearing are oil-impregnated in the same way as the front and middle crank-shaft bearings, and they are held together by two 3/8" diameter bolts through top and bottom plates, item 80b, each of 1½" x 1¼" mild steel.

    The top plate is drilled and threaded to take a screw-type grease-cup, item 80c.

    A ½" diameter mild steel rod, item 80d, is welded to the bottom plate at a 5° angle to the bearing face to match the crankshaft inclination and triangular ¼" thick reinforcement webs are welded each side of the rod to provide adequate strength.

    The rod, item 80d, is welded inside a piece of 5/8" outer diameter conduit pipe, which then connects with the crank rod, item 79a, via the coupling, item 79b.

The erected wind-pump in action.

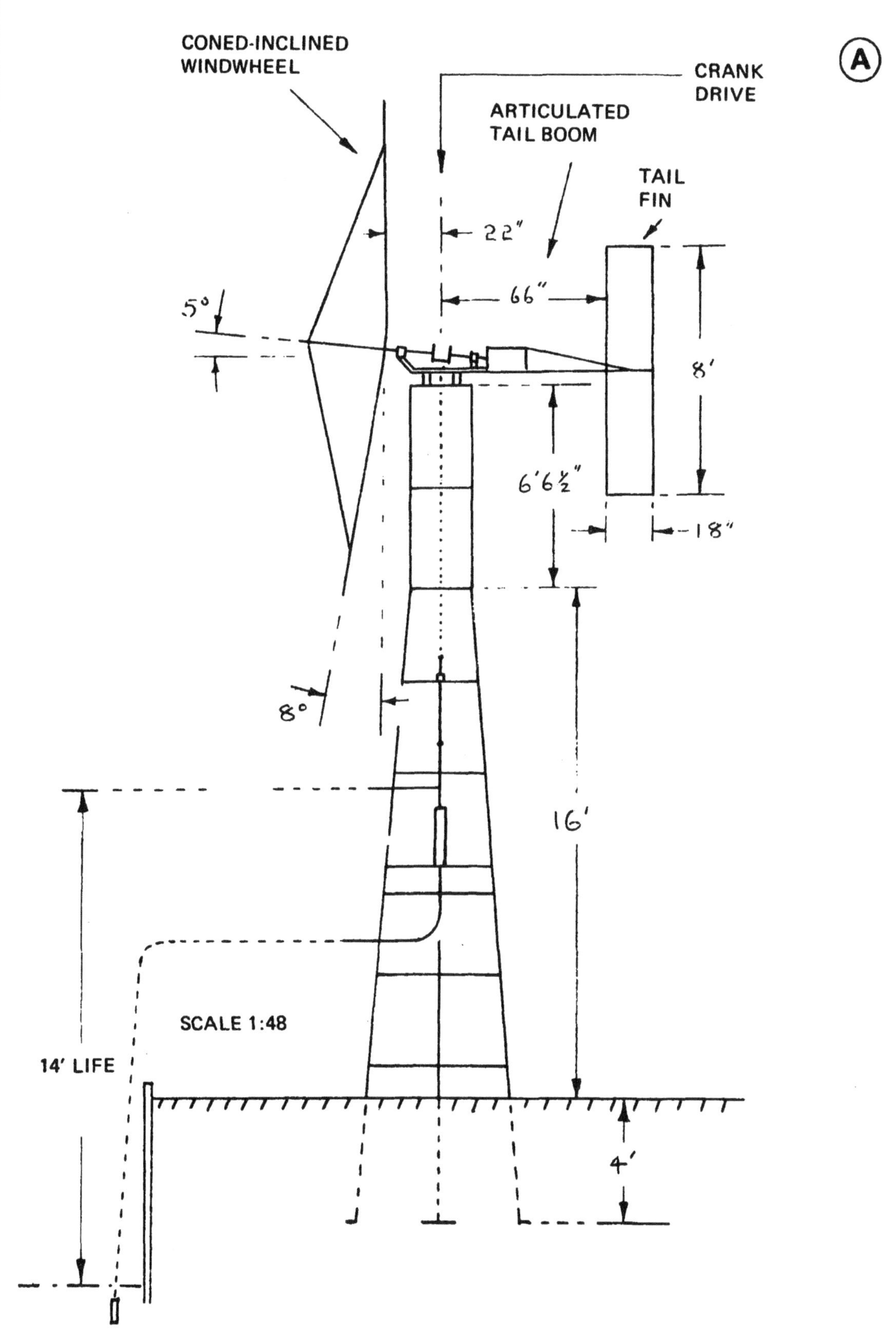
CONED-INCLINED WINDWHEEL
CRANK DRIVE
A
ARTICULATED TAIL BOOM
TAIL FIN
22"
66"
5°
8'
6'6½"
18"
8°
16'
SCALE 1:48
14' LIFE
4'

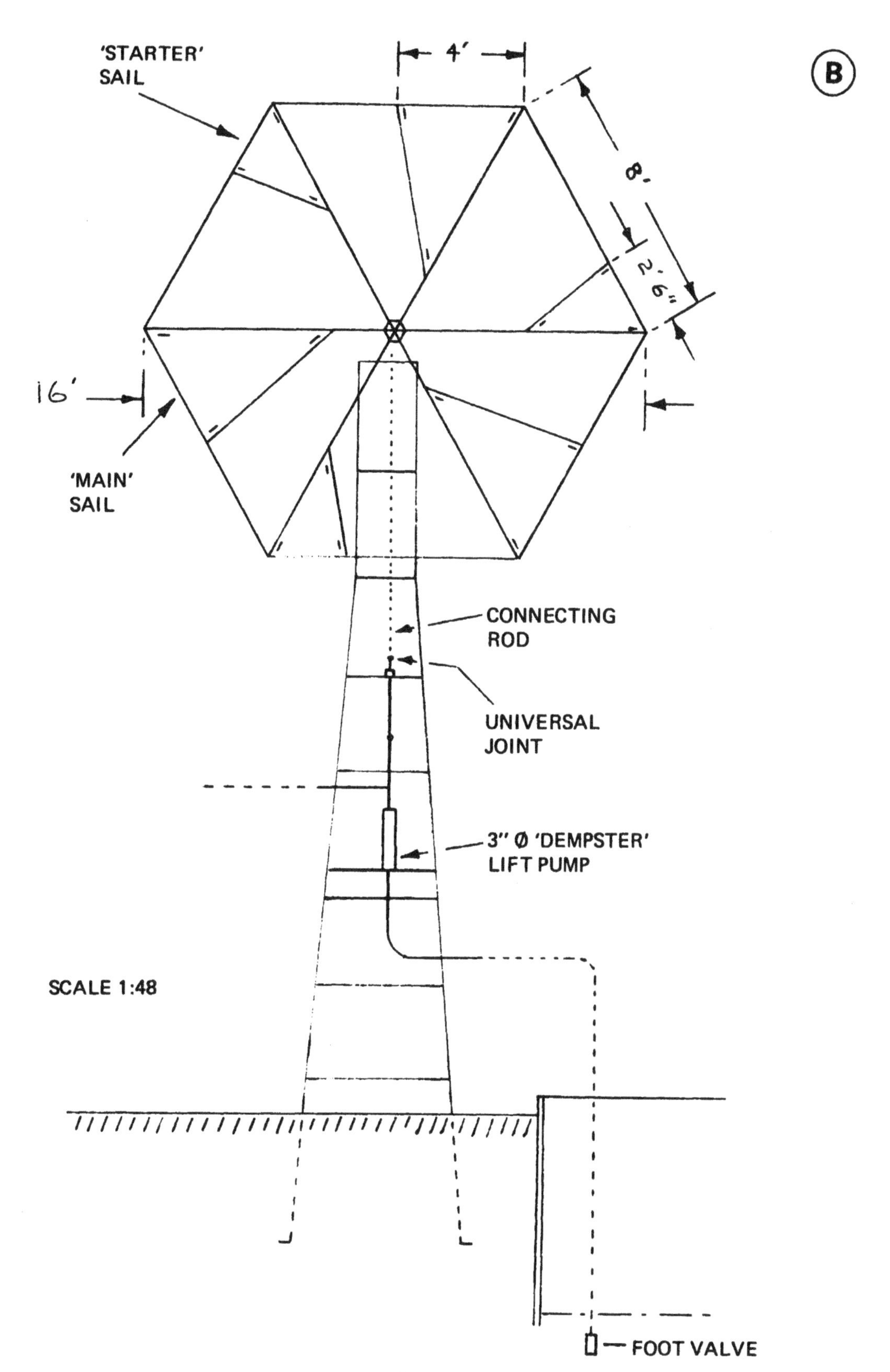
B
'STARTER'
SAIL
4'
8'
2'6"
16'
'MAIN'
SAIL
CONNECTING
ROD
UNIVERSAL
JOINT
3" Ø 'DEMPSTER'
LIFT PUMP
SCALE 1:48
FOOT VALVE

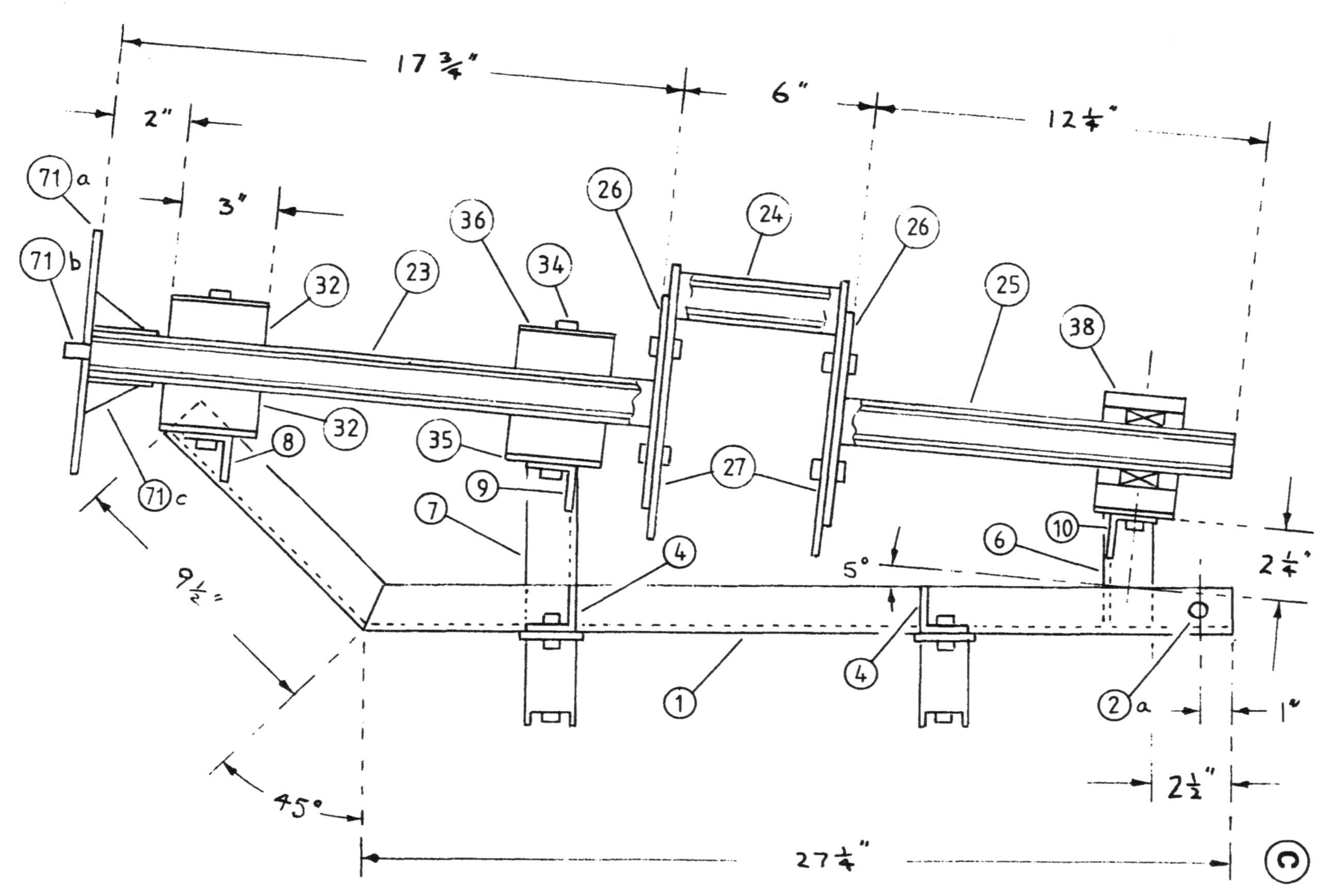
17¾"
6"
12¼"
2"
3"
71a
71b
32
23
36
34
26
24
26
25
38
32
8
35
9
7
27
4
5°
10
6
2¼"
71c
9½"
4
1
2a
1"
2½"
45°
27¼"
C

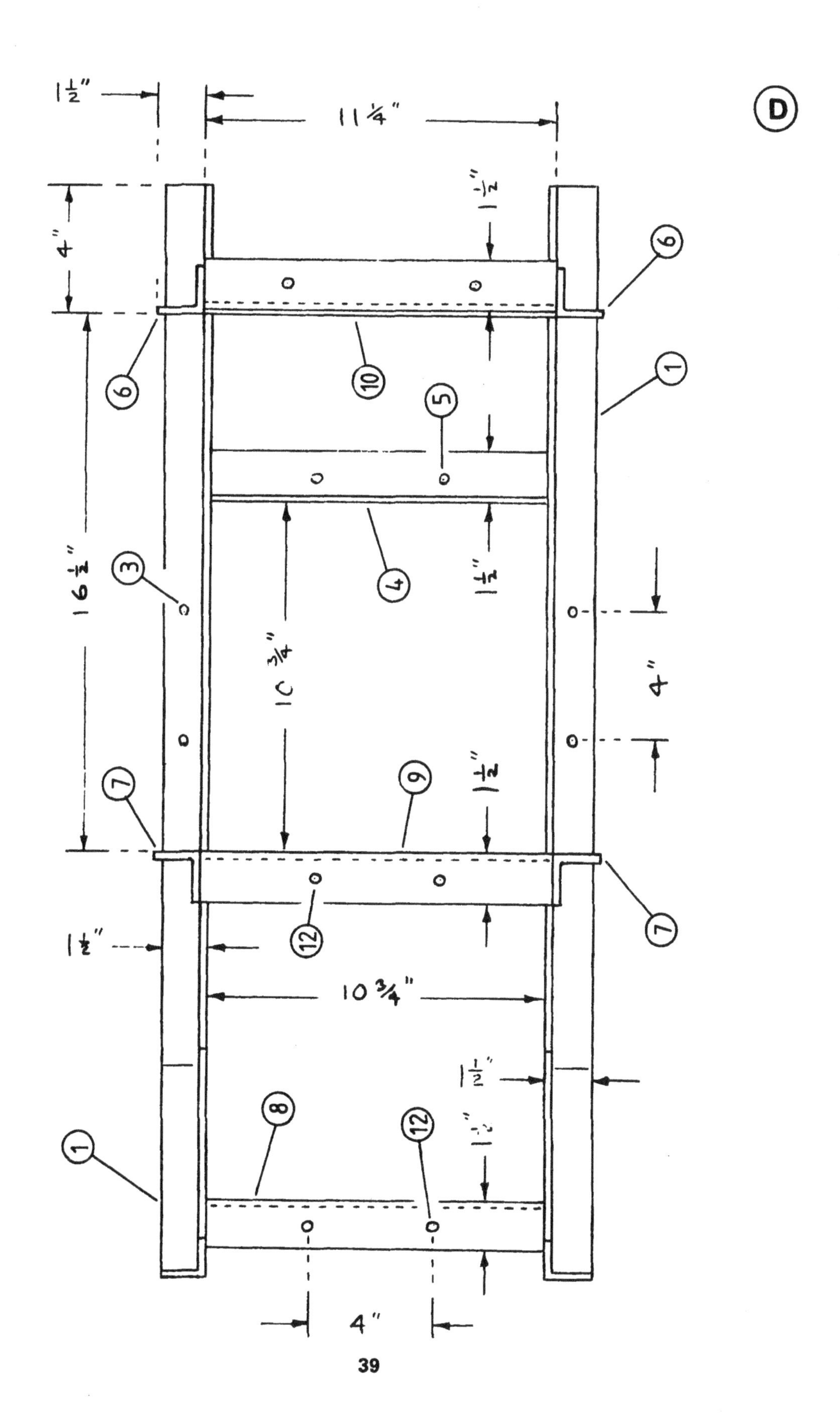
D
1½"
11¼"
4"
16½"
10¾"
1½"
10¾"
1½"
4"
4"
1
3
4
5
6
7
8
9
10
12

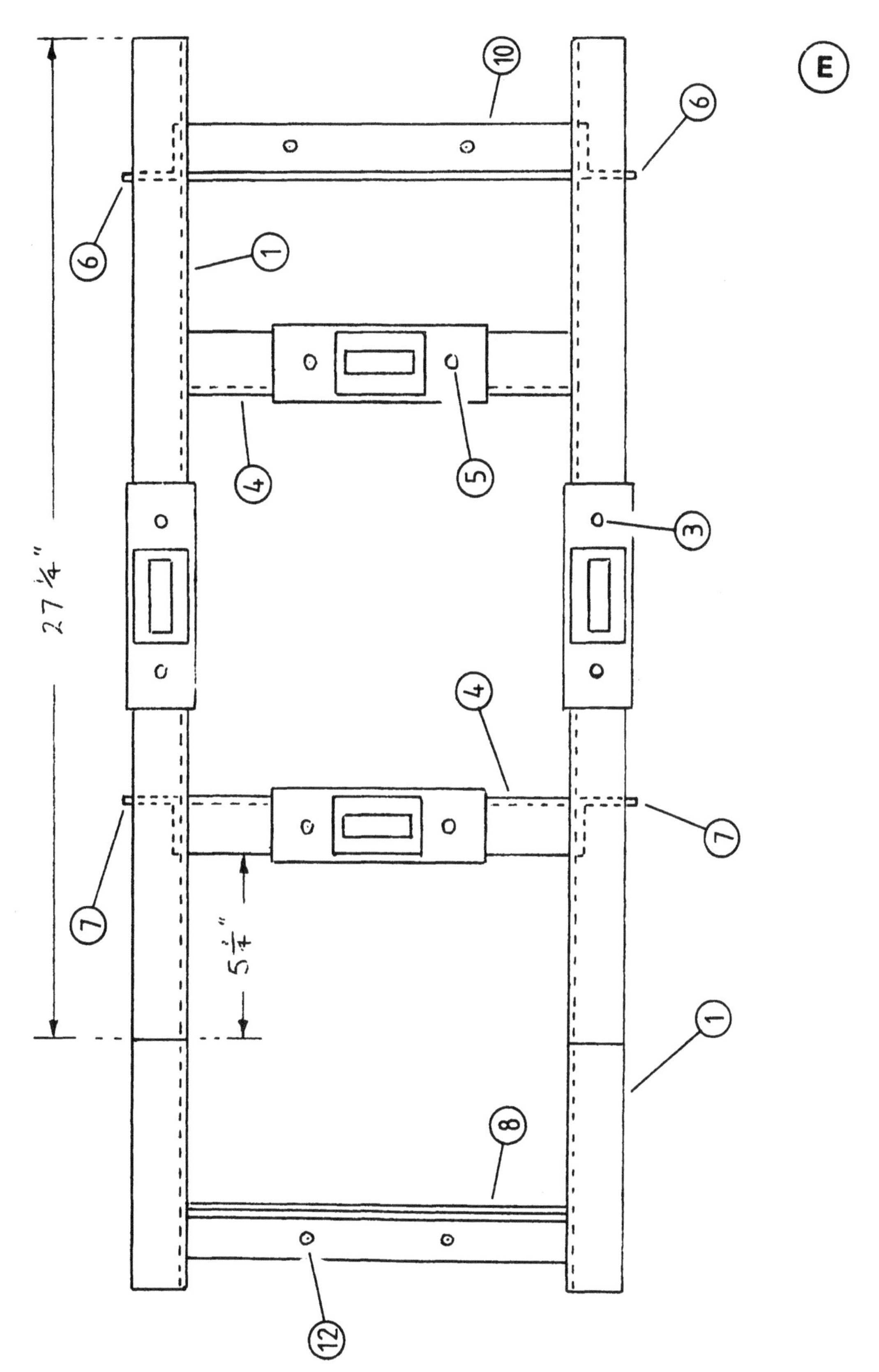
E
27 1/4"
5 1/4"
1
3
4
5
6
7
8
10
12

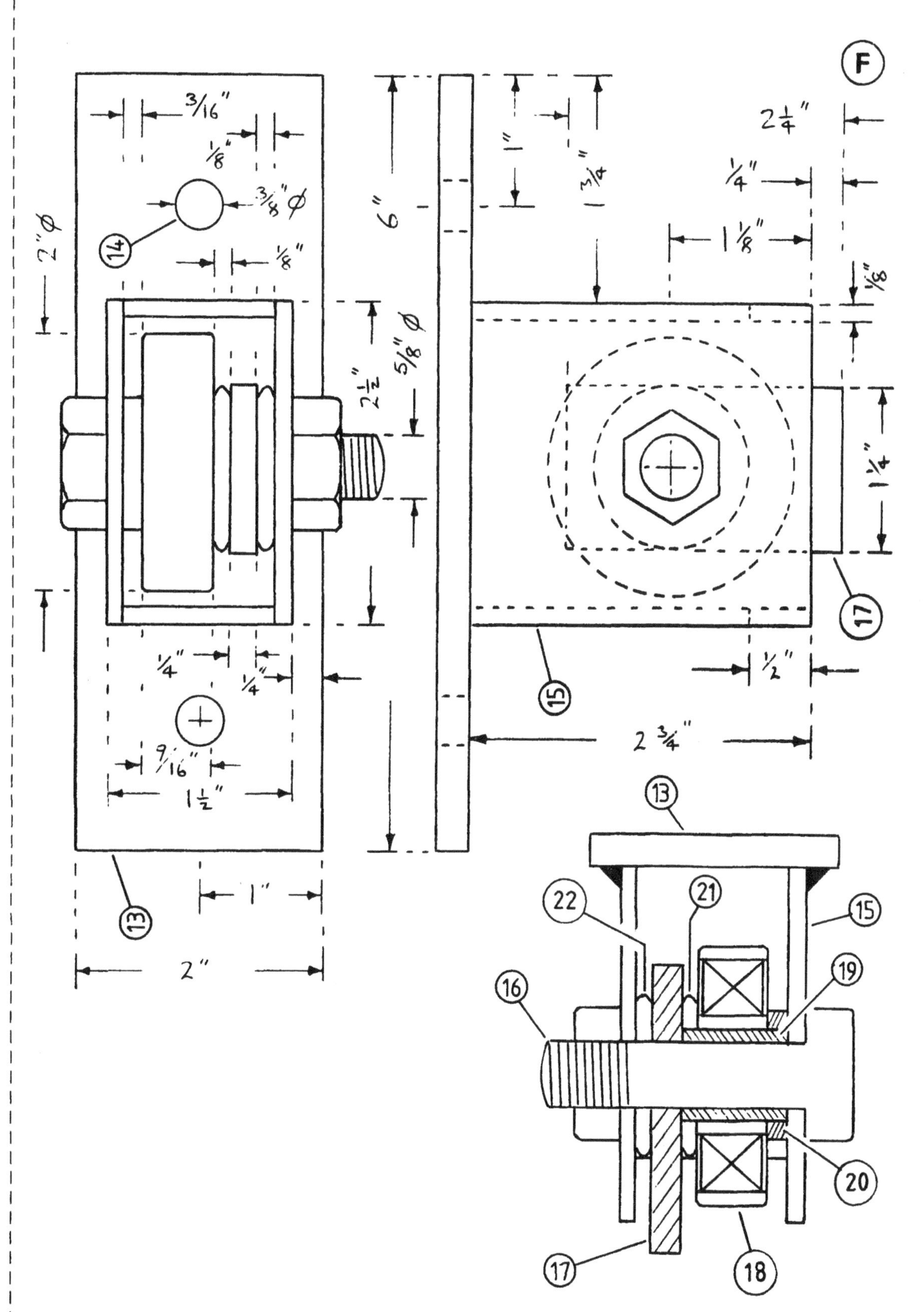
F
3/16"
1/8"
3/8" Ø
2" Ø
14
1/8"
6"
1"
1 3/4"
2 1/4"
1/4"
1 1/8"
1/8"
5/8" Ø
2 1/2"
1 1/4"
17
15
1/2"
1/4"
1/4"
9/16"
1 1/2"
2 3/4"
13
1"
2"
13
22
21
15
19
16
20
17
18

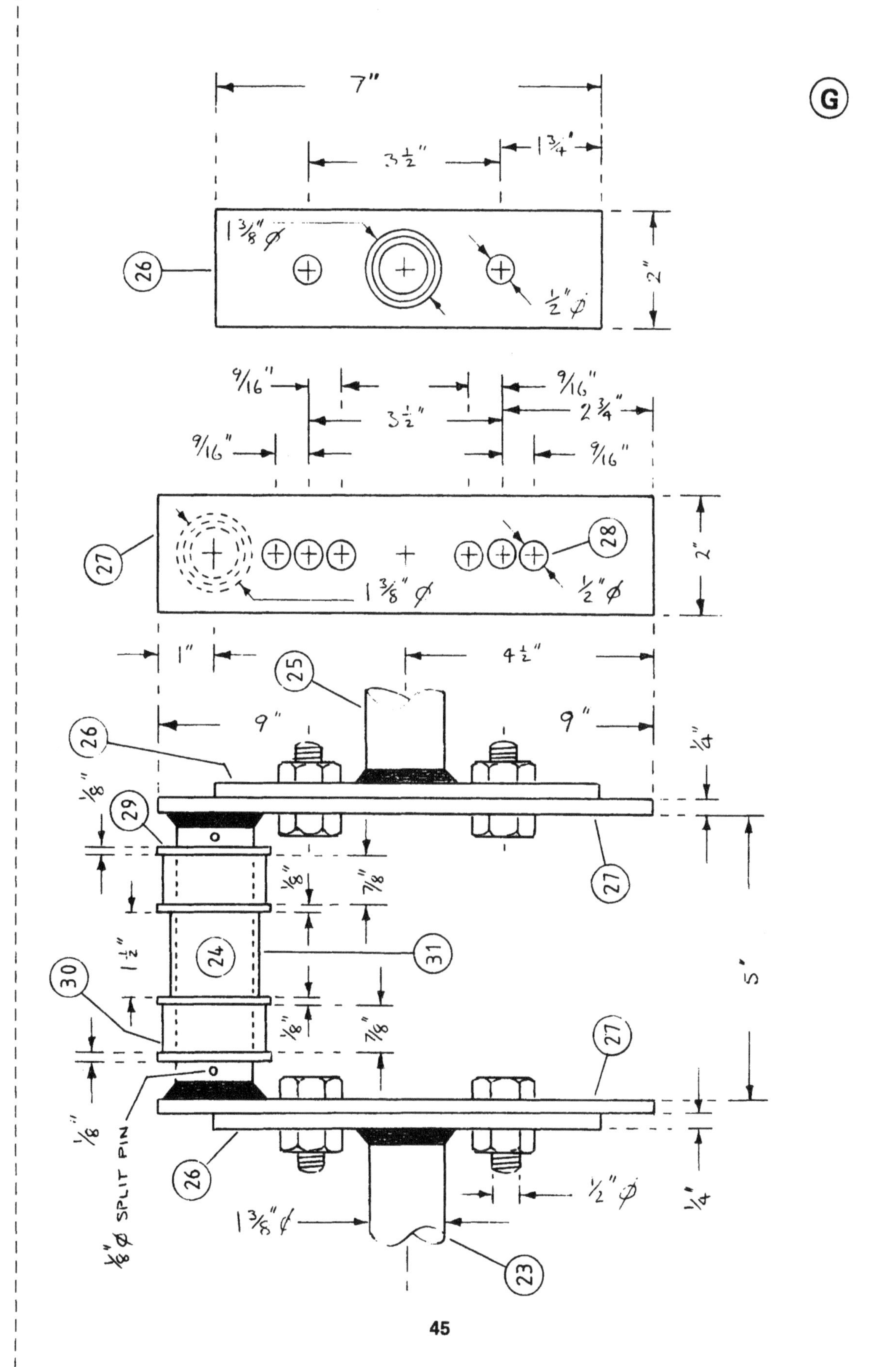

G
7"
3½"
1¾"
1⅜" ⌀
½" ⌀
2"
26
9/16"
3½"
2¾"
9/16"
27
28
1⅜" ⌀
½" ⌀
2"
1"
4½"
25
9"
9"
26
¼"
⅛"
29
⅛"
⅞"
1½"
24
31
30
⅛"
⅞"
5"
27
27
⅛"
⅛" ⌀ SPLIT PIN
26
½" ⌀
¼"
1⅜" ⌀
23

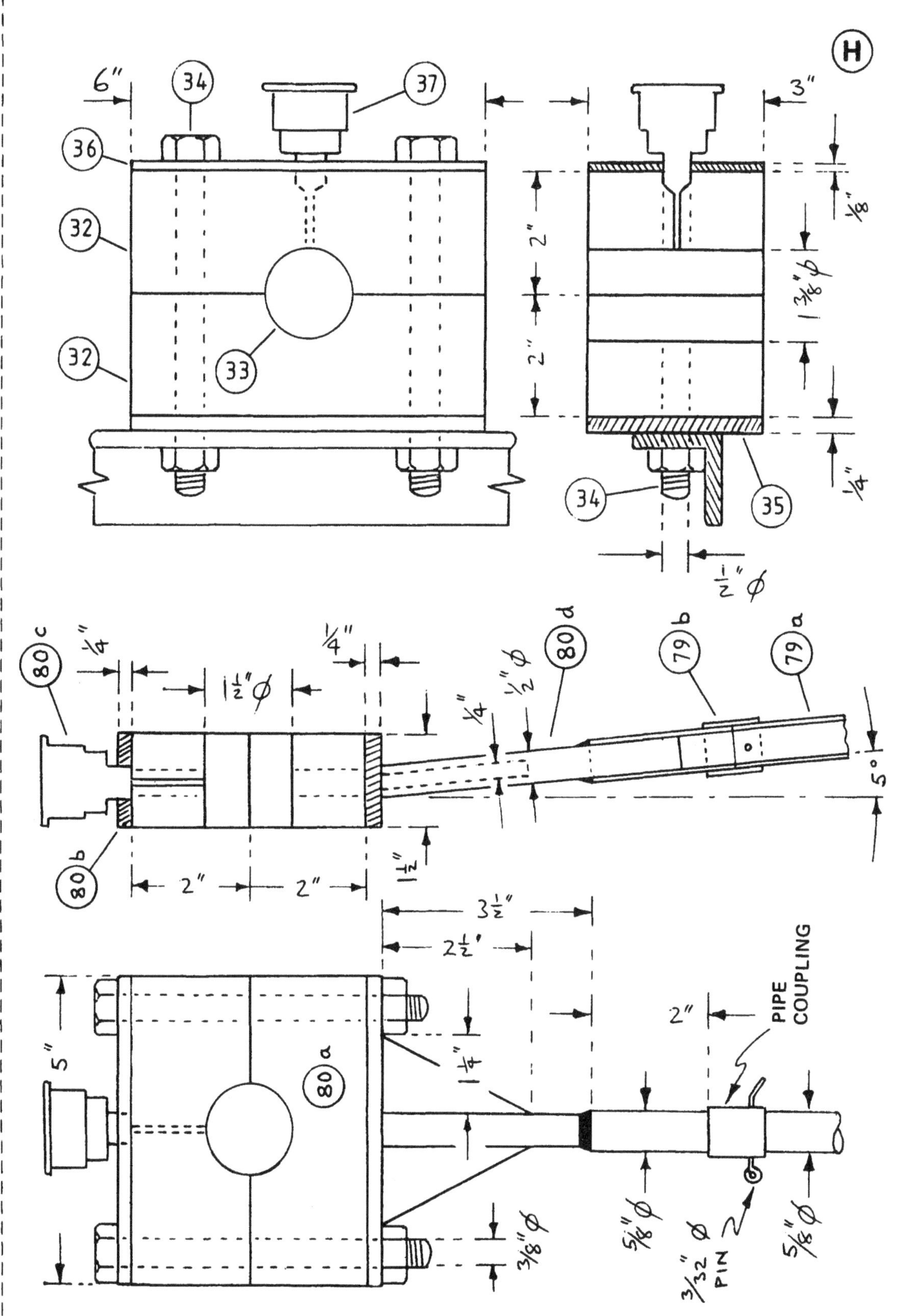
H
34
37
36
32
32
33
34
35
6"
3"
2"
2"
1/8"
1/4"
1 3/16" ⌀
1/2" ⌀
80 c
80 b
80 d
79 b
79 a
80 a
1/4"
1/4"
1 1/2" ⌀
1/2" ⌀
1 1/2"
2"
2"
5°
3 1/2"
2 1/2"
2"
5"
1 1/4"
3/8" ⌀
5/8" ⌀
5/16" ⌀
3/32" ⌀
PIN
PIPE
COUPLING

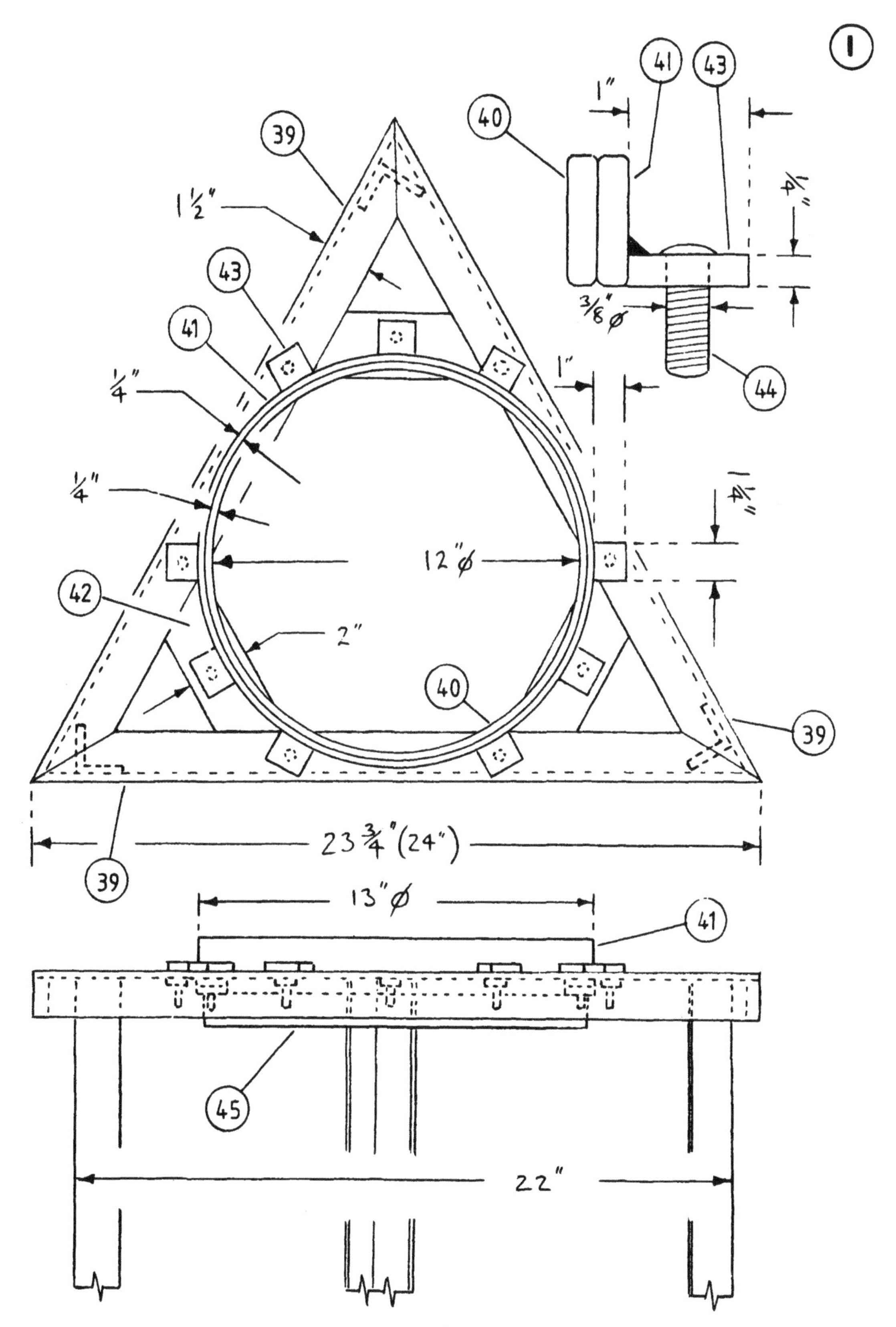

1
41
43
40
39
1"
1 1/2"
1/4"
43
41
3/8" ø
44
1/4"
1"
12" ø
42
2"
40
39
23 3/4" (24")
39
13" ø
41
45
22"

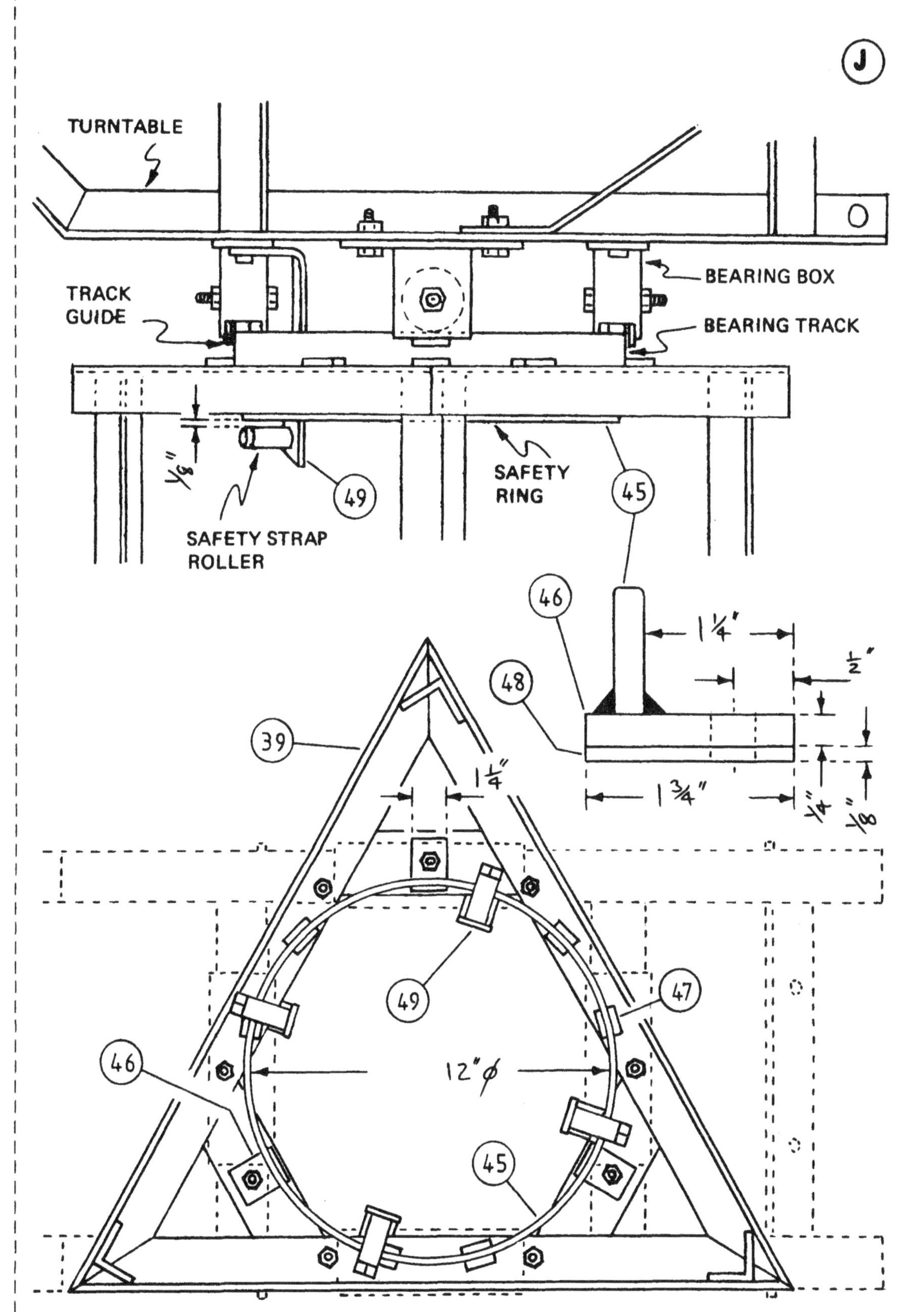
J
TURNTABLE
TRACK GUIDE
BEARING BOX
BEARING TRACK
SAFETY STRAP ROLLER
SAFETY RING
49
45
46
48
39
47
12"ϕ

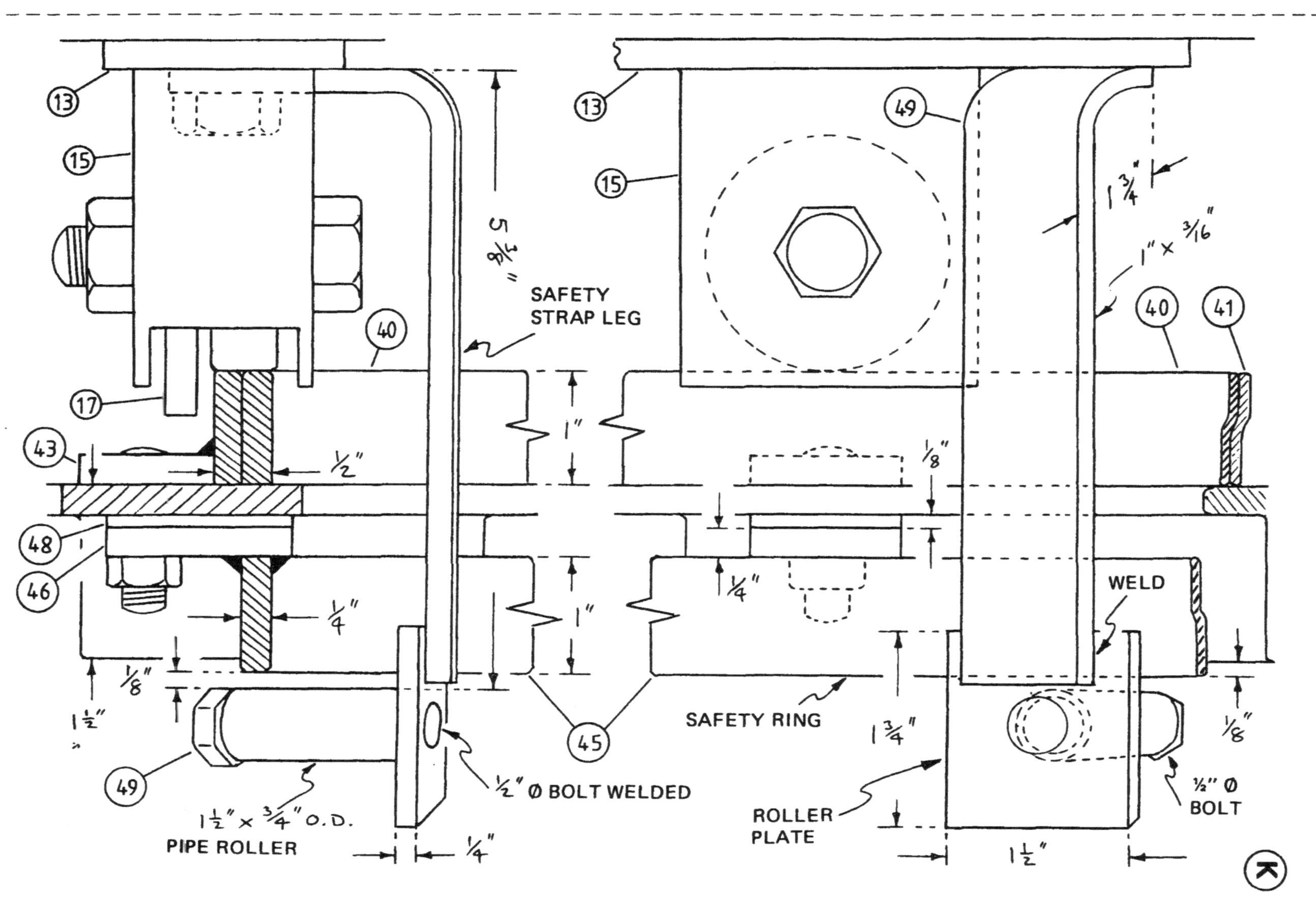
13
15
5 3/8"
SAFETY STRAP LEG
40
17
43
1/2"
48
46
1/4"
1/8"
1 1/2"
49
1 1/2" x 3/4" O.D.
PIPE ROLLER
1/4"
1/2" Ø BOLT WELDED
1"
1"
45
13
49
15
1 3/4"
1" x 3/16"
40
41
1/8"
1/4"
WELD
SAFETY RING
1 3/4"
ROLLER PLATE
1 1/2"
1/8"
1/2" Ø BOLT
K

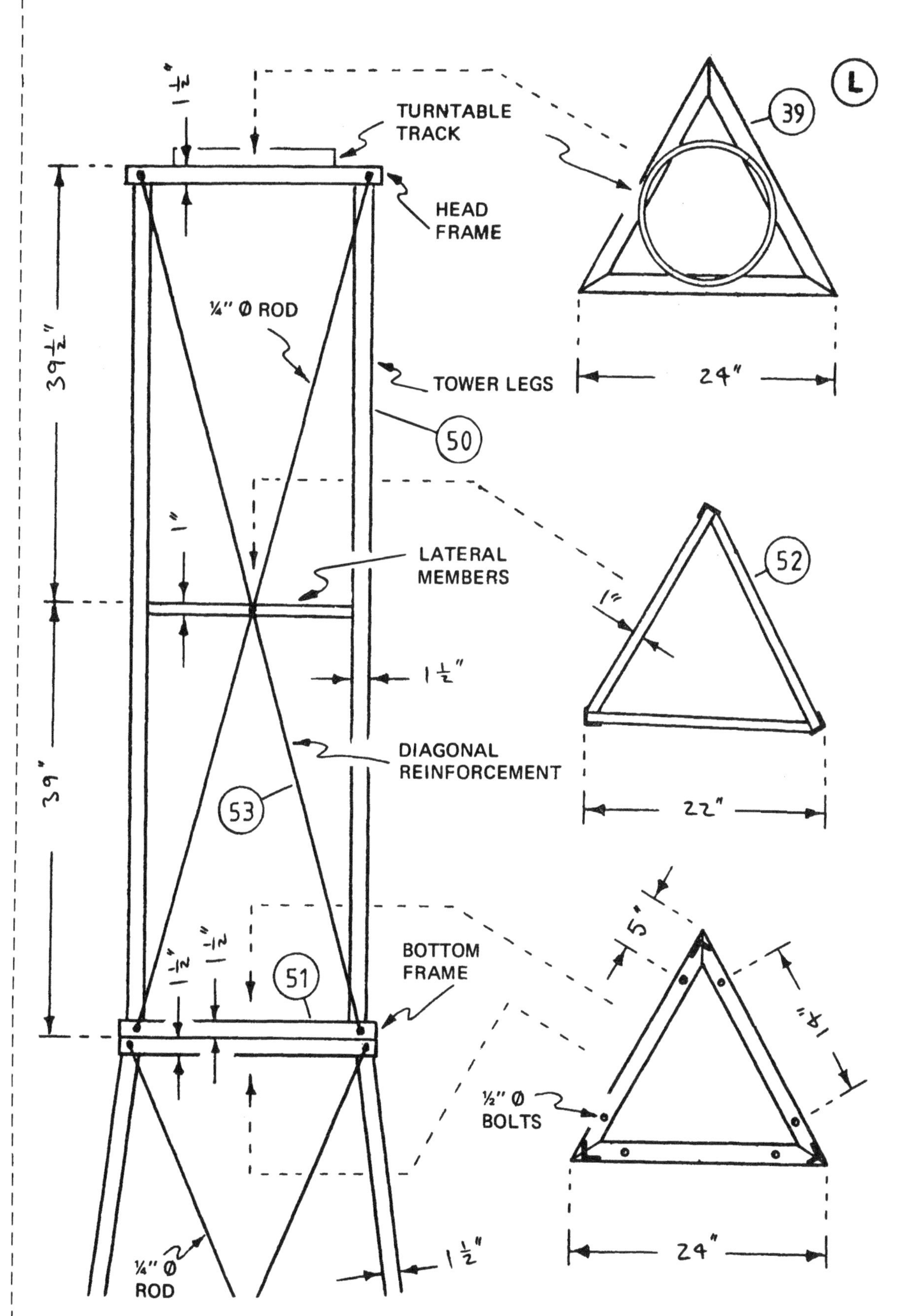

TURNTABLE TRACK
HEAD FRAME
¼" Ø ROD
TOWER LEGS
LATERAL MEMBERS
DIAGONAL REINFORCEMENT
BOTTOM FRAME
½" Ø BOLTS
¼" Ø ROD
L
39
50
52
53
51
24"
22"
24"
39½"
39"
1½"
1"
1½"
1"
5"
18"
1½"

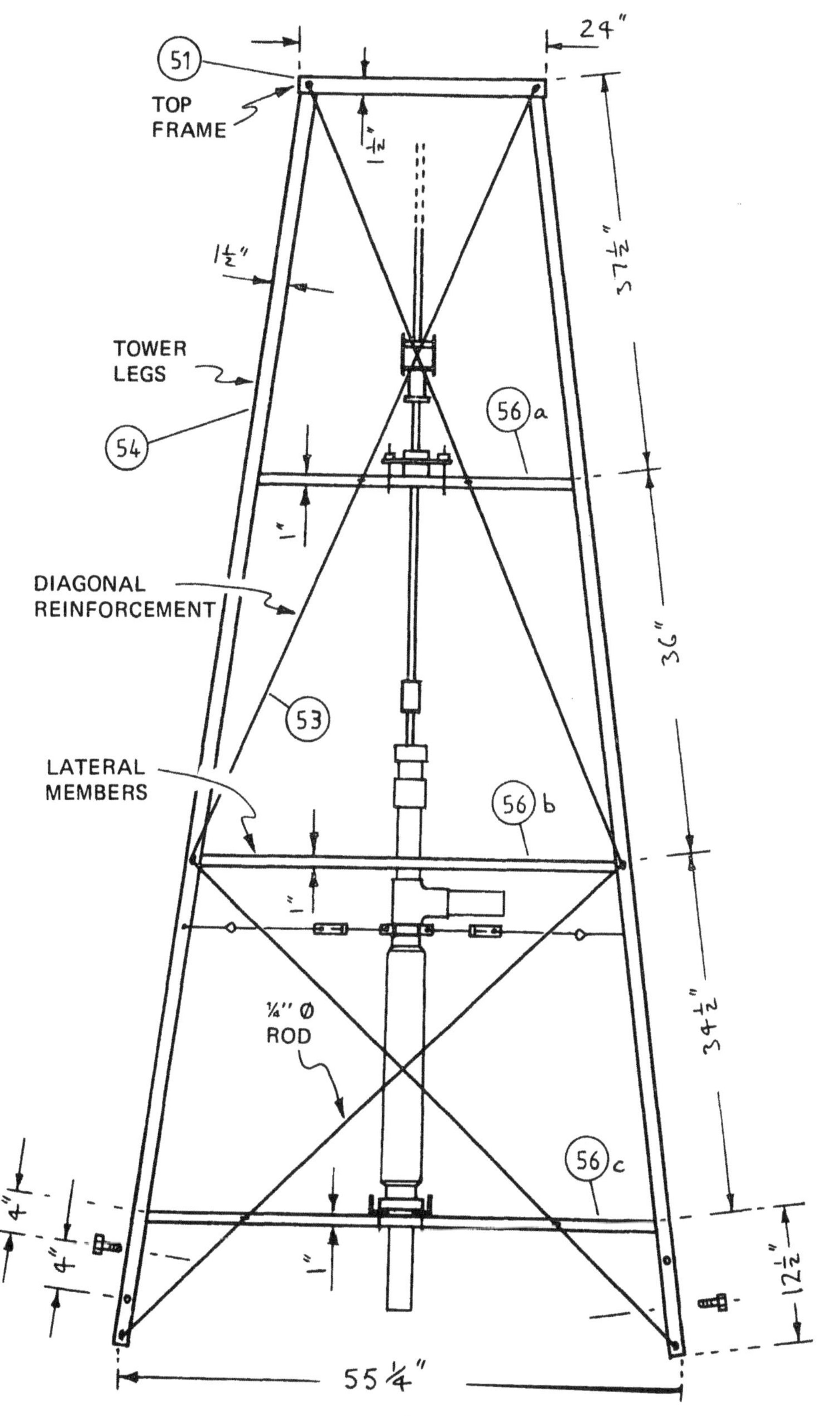

24"
51
TOP
FRAME
1/2"
1 1/2"
37 1/2"
TOWER
LEGS
54
56 a
1"
DIAGONAL
REINFORCEMENT
36"
53
LATERAL
MEMBERS
56 b
1"
34 1/2"
1/4" Ø
ROD
56 c
4"
4"
1"
12 1/2"
55 1/4"

M

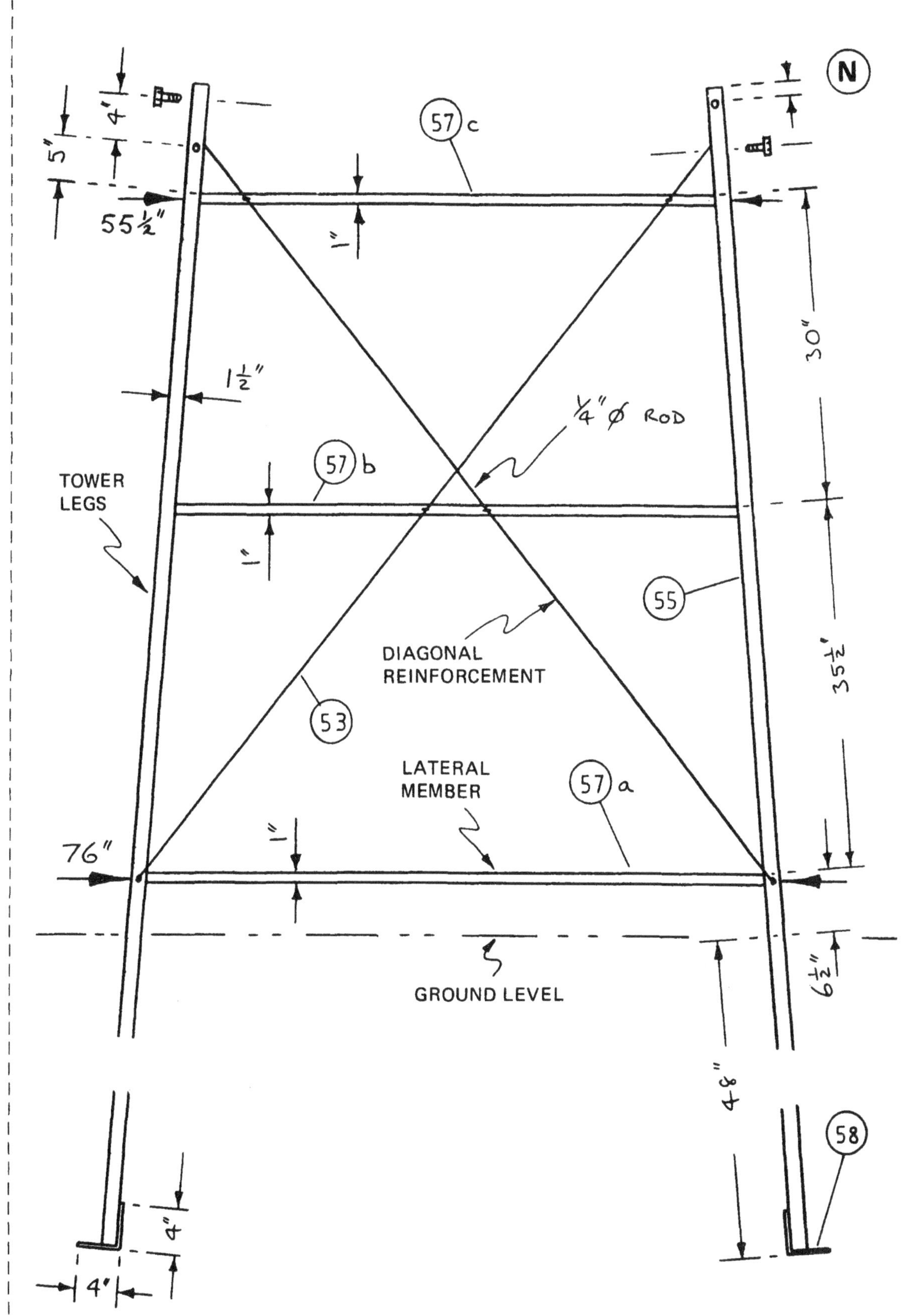

N
4"
5"
55½"
57 c
1"
30"
1½"
¼" ⌀ ROD
57 b
1"
TOWER
LEGS
55
35½"
DIAGONAL
REINFORCEMENT
53
LATERAL
MEMBER
57 a
1"
76"
GROUND LEVEL
6½"
48"
58
4"
4"

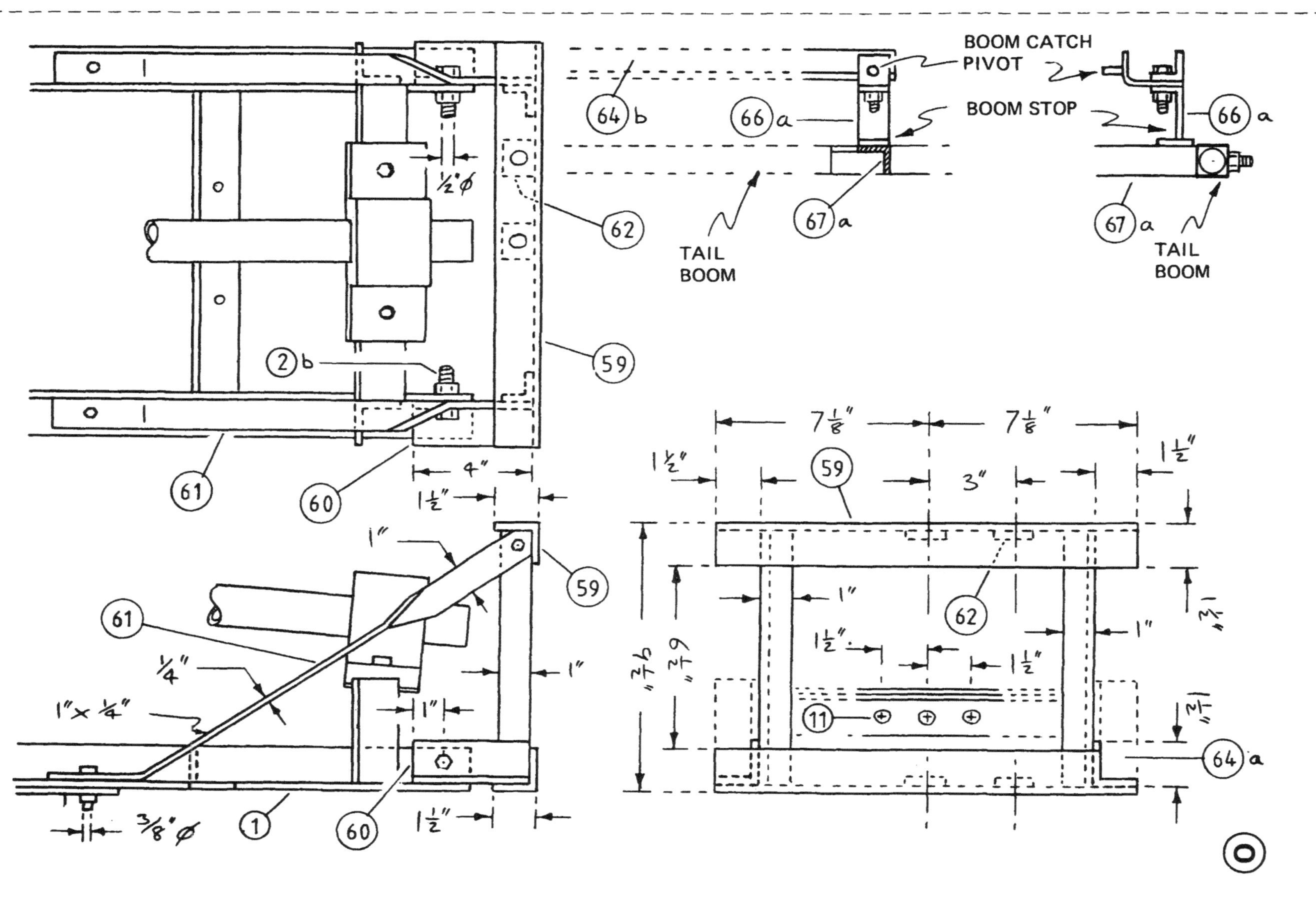
BOOM CATCH PIVOT
BOOM STOP
TAIL BOOM
TAIL BOOM
64b
66a
67a
66a
67a
62
59
2b
61
60
59
61
1
60
59
62
11
64a
O
1/2"ϕ
4"
1 1/2"
1"
1/4"
1"x 1/4"
3/8"ϕ
7 1/8"
7 1/8"
3"
1 1/2"
9 1/2"
6 1/2"

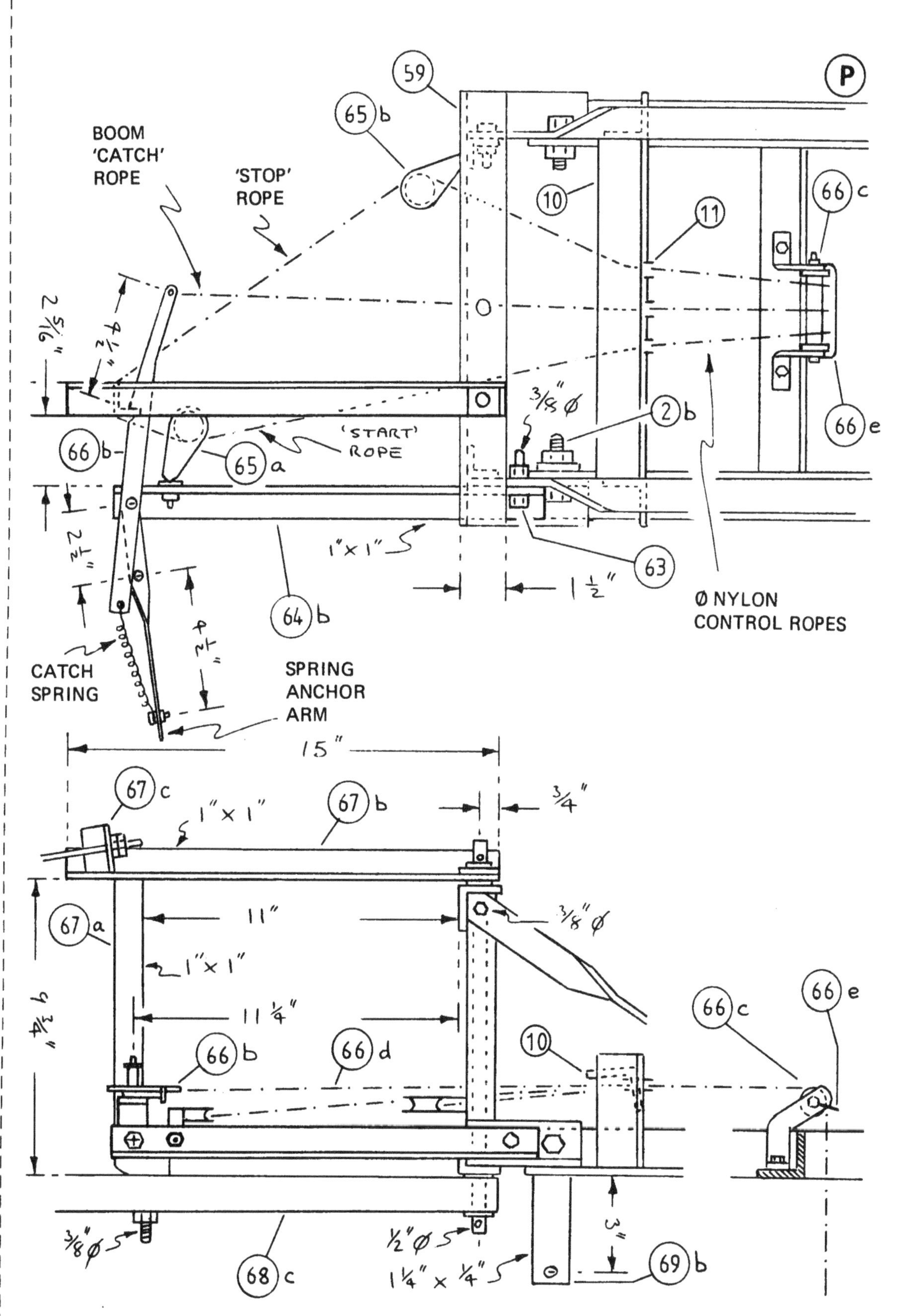
P
59
65 b
BOOM
'CATCH'
ROPE
'STOP'
ROPE
10
11
66 c
2 5/16"
4 1/2"
3/8" Ø
2 b
66 e
66 b
'START'
ROPE
65 a
2 1/4"
1" x 1"
63
1 1/2"
64 b
4 1/4"
Ø NYLON
CONTROL ROPES
CATCH
SPRING
SPRING
ANCHOR
ARM
15"
67 c
1" x 1"
67 b
3/4"
67 a
11"
3/8" Ø
1" x 1"
4 1/2"
11 1/4"
66 c
66 e
66 b
66 d
10
3/8" Ø
1/2" Ø
3"
68 c
1 1/4" x 1/4"
69 b

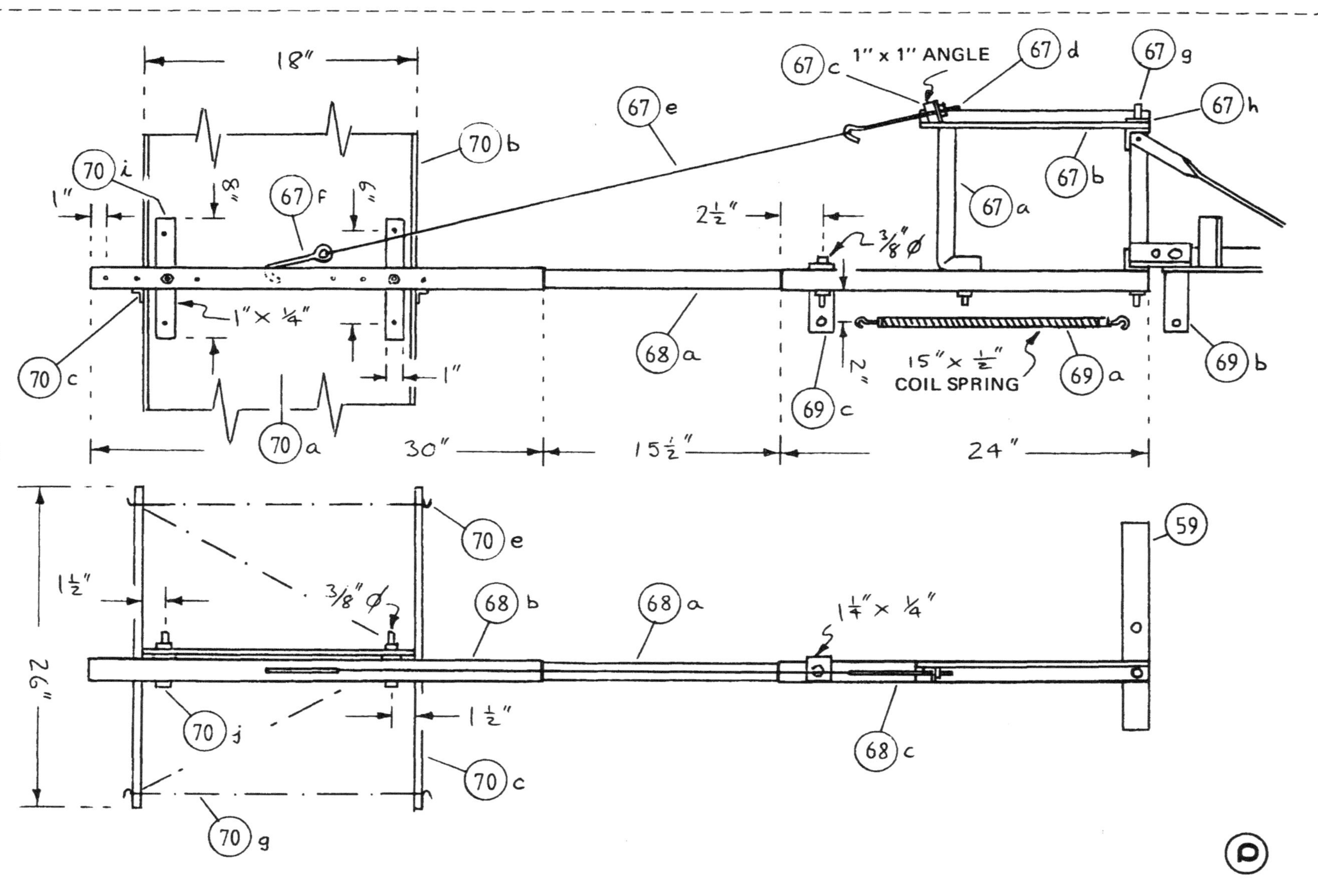

18"
67 c
1" x 1" ANGLE
67 d
67 g
67 h
67 e
70 b
67 b
70 i
1"
8"
67 f
6"
67 a
2 1/2"
3/8" Ø
1" x 1/4"
68 a
2"
15" x 1/2"
COIL SPRING
69 a
69 b
70 c
1"
69 c
70 a
30"
15 1/2"
24"
70 e
59
1 1/2"
3/8" Ø
68 b
68 a
1 1/4" x 1/4"
26"
1 1/2"
70 j
68 c
70 c
70 g
Q

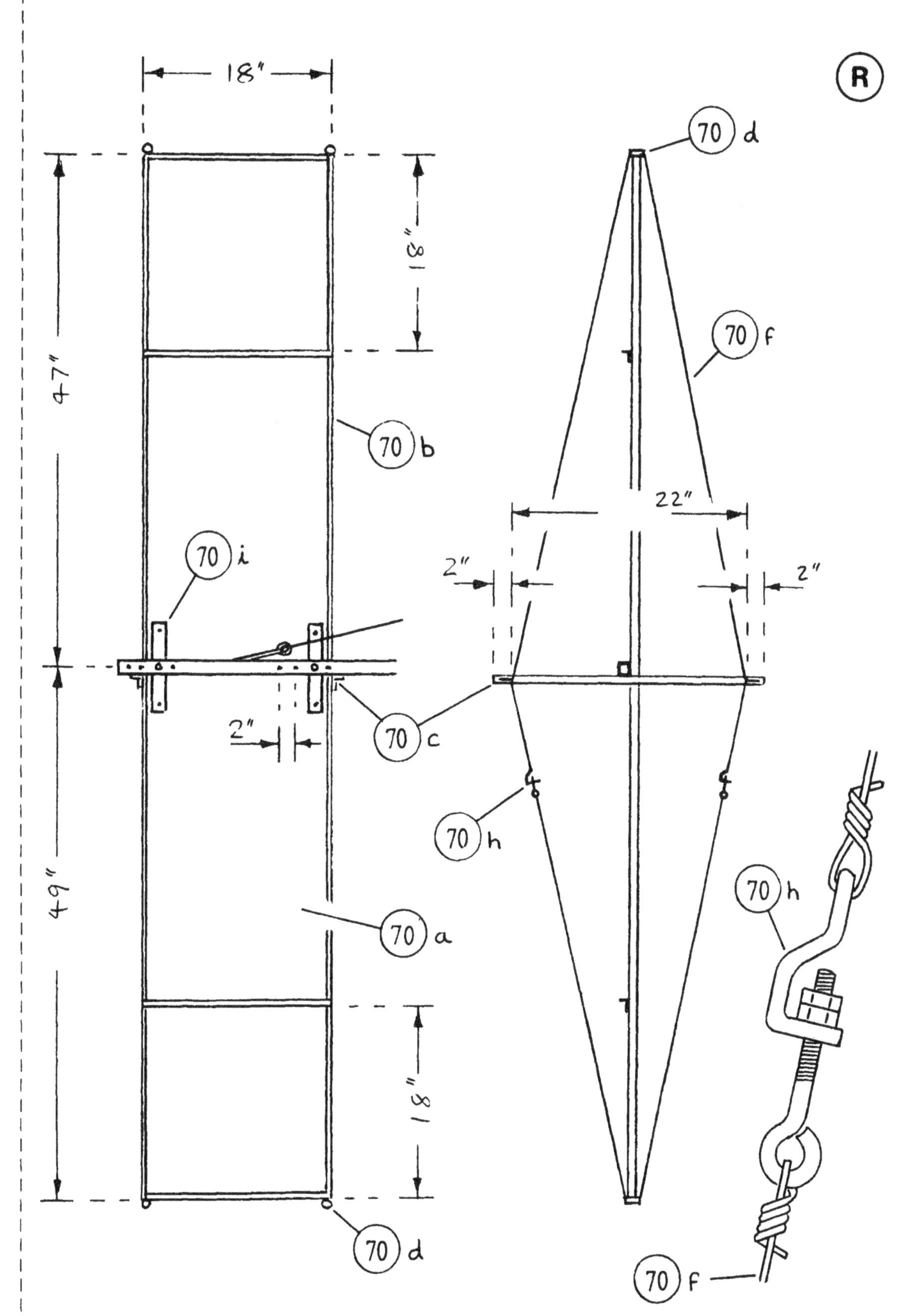

R
18"
47"
18"
49"
18"
2"
22"
2"
2"
70 a
70 b
70 c
70 d
70 d
70 f
70 f
70 h
70 h
70 i

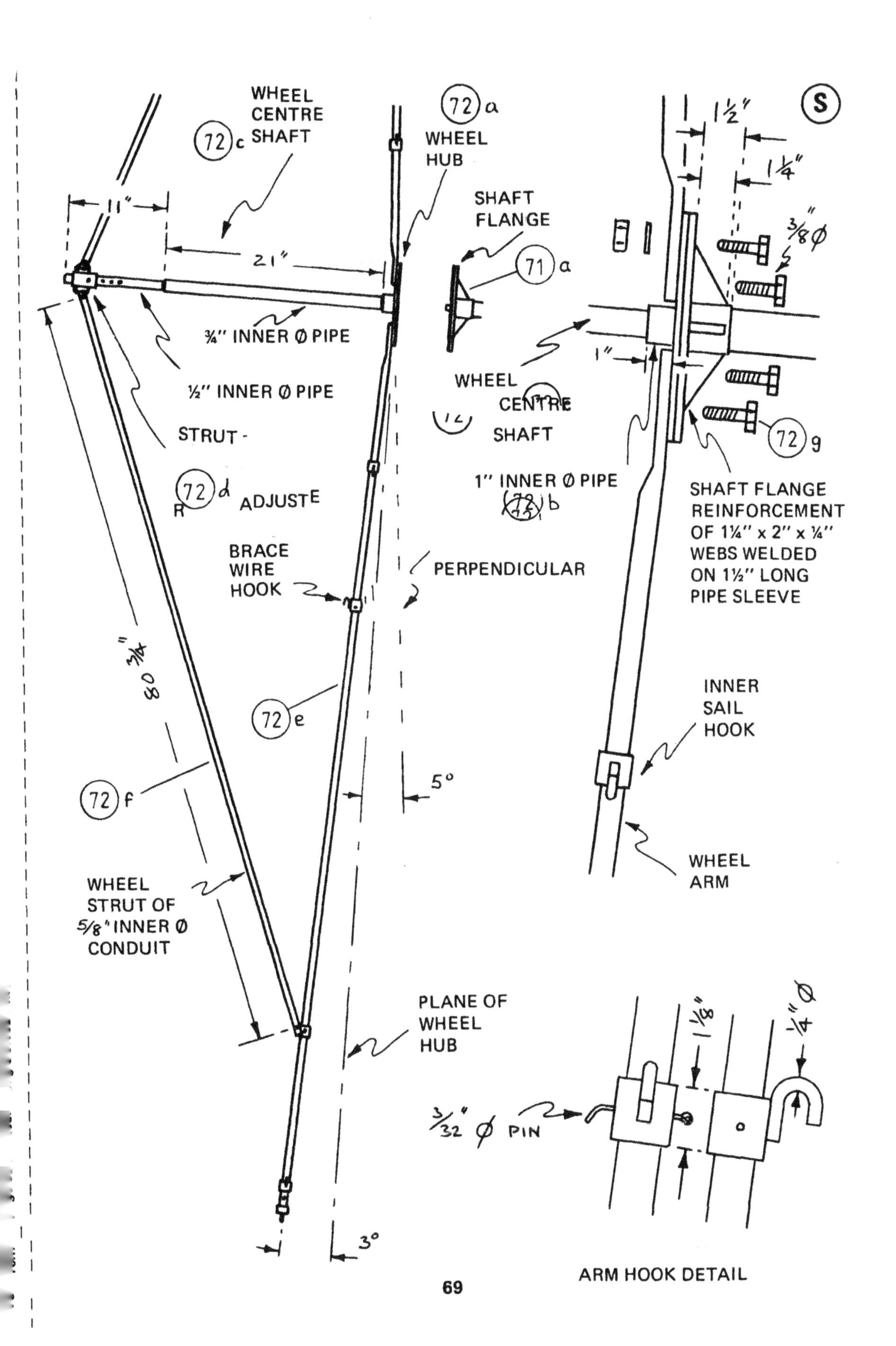
WHEEL
CENTRE
SHAFT
72 c
72 a
WHEEL
HUB
S
1½"
1¼"
11"
21"
SHAFT
FLANGE
71 a
3/8 Ø
¾" INNER Ø PIPE
½" INNER Ø PIPE
WHEEL
CENTRE
SHAFT
1"
STRUT -
ADJUSTER
72 d
1" INNER Ø PIPE
72 b
72 g
SHAFT FLANGE
REINFORCEMENT
OF 1¼" x 2" x ¼"
WEBS WELDED
ON 1½" LONG
PIPE SLEEVE
BRACE
WIRE
HOOK
PERPENDICULAR
80 ¾"
72 e
INNER
SAIL
HOOK
72 f
5°
WHEEL
ARM
WHEEL
STRUT OF
5/8" INNER Ø
CONDUIT
PLANE OF
WHEEL
HUB
1 1/8"
¼" Ø
3/32" Ø PIN
3°
ARM HOOK DETAIL

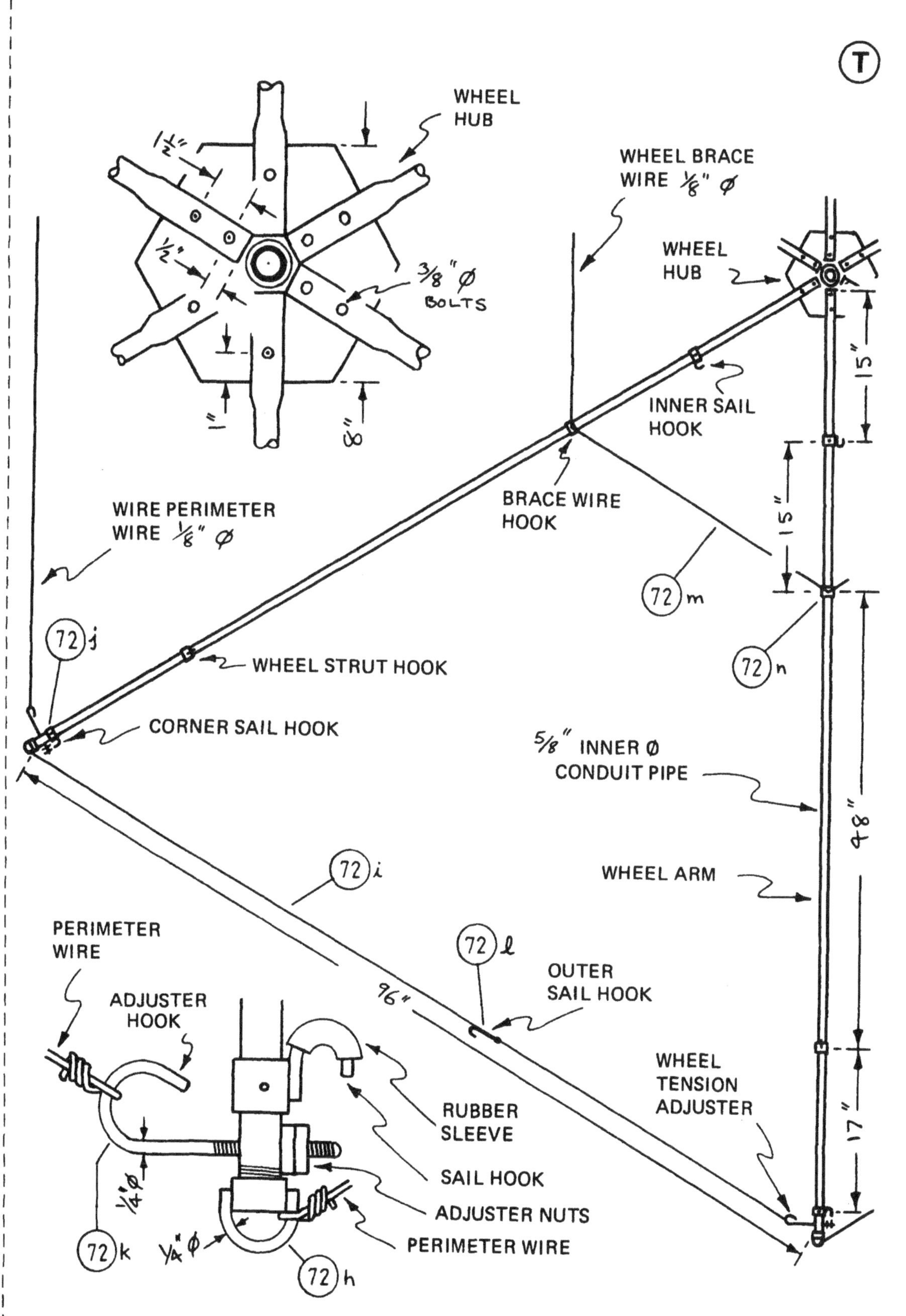
T
WHEEL
HUB
1½"
½"
3/8" Ø
BOLTS
1"
8"
WHEEL BRACE
WIRE ⅛" Ø
WHEEL
HUB
15"
INNER SAIL
HOOK
15"
BRACE WIRE
HOOK
WIRE PERIMETER
WIRE ⅛" Ø
72 m
72 j
WHEEL STRUT HOOK
72 n
CORNER SAIL HOOK
5/8" INNER Ø
CONDUIT PIPE
48"
72 i
WHEEL ARM
PERIMETER
WIRE
72 l
ADJUSTER
HOOK
96"
OUTER
SAIL HOOK
WHEEL
TENSION
ADJUSTER
RUBBER
SLEEVE
17"
¼" Ø
SAIL HOOK
ADJUSTER NUTS
72 k
¼" Ø
PERIMETER WIRE
72 h

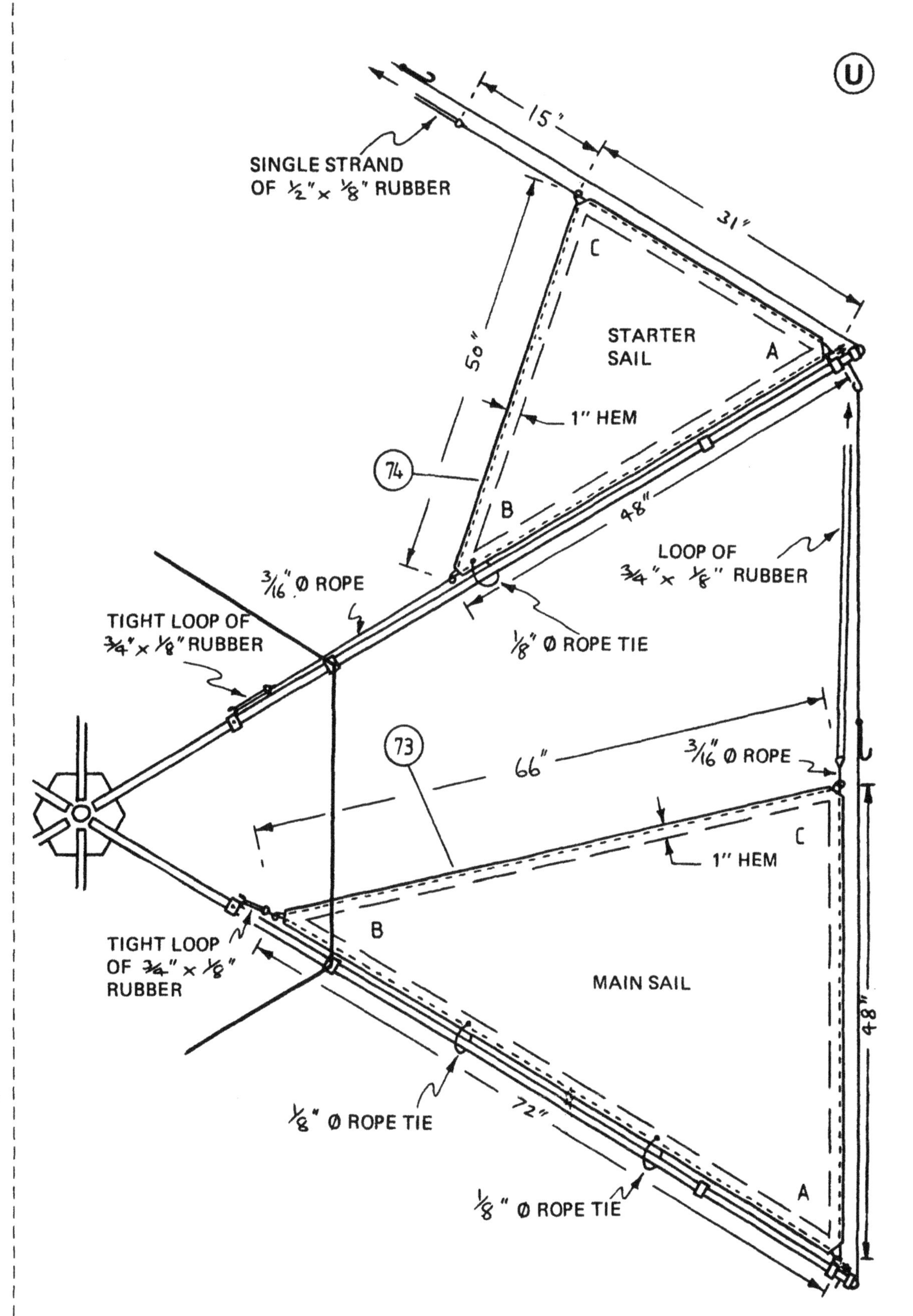
U
15"
SINGLE STRAND
OF ½" × ⅛" RUBBER
31"
50"
C
STARTER
SAIL
A
1" HEM
74
B
48"
LOOP OF
¾" × ⅛" RUBBER
3/16" Ø ROPE
TIGHT LOOP OF
¾" × ⅛" RUBBER
⅛" Ø ROPE TIE
73
66"
3/16" Ø ROPE
C
1" HEM
B
TIGHT LOOP
OF ¾" × ⅛"
RUBBER
MAIN SAIL
48"
⅛" Ø ROPE TIE
72"
A
⅛" Ø ROPE TIE

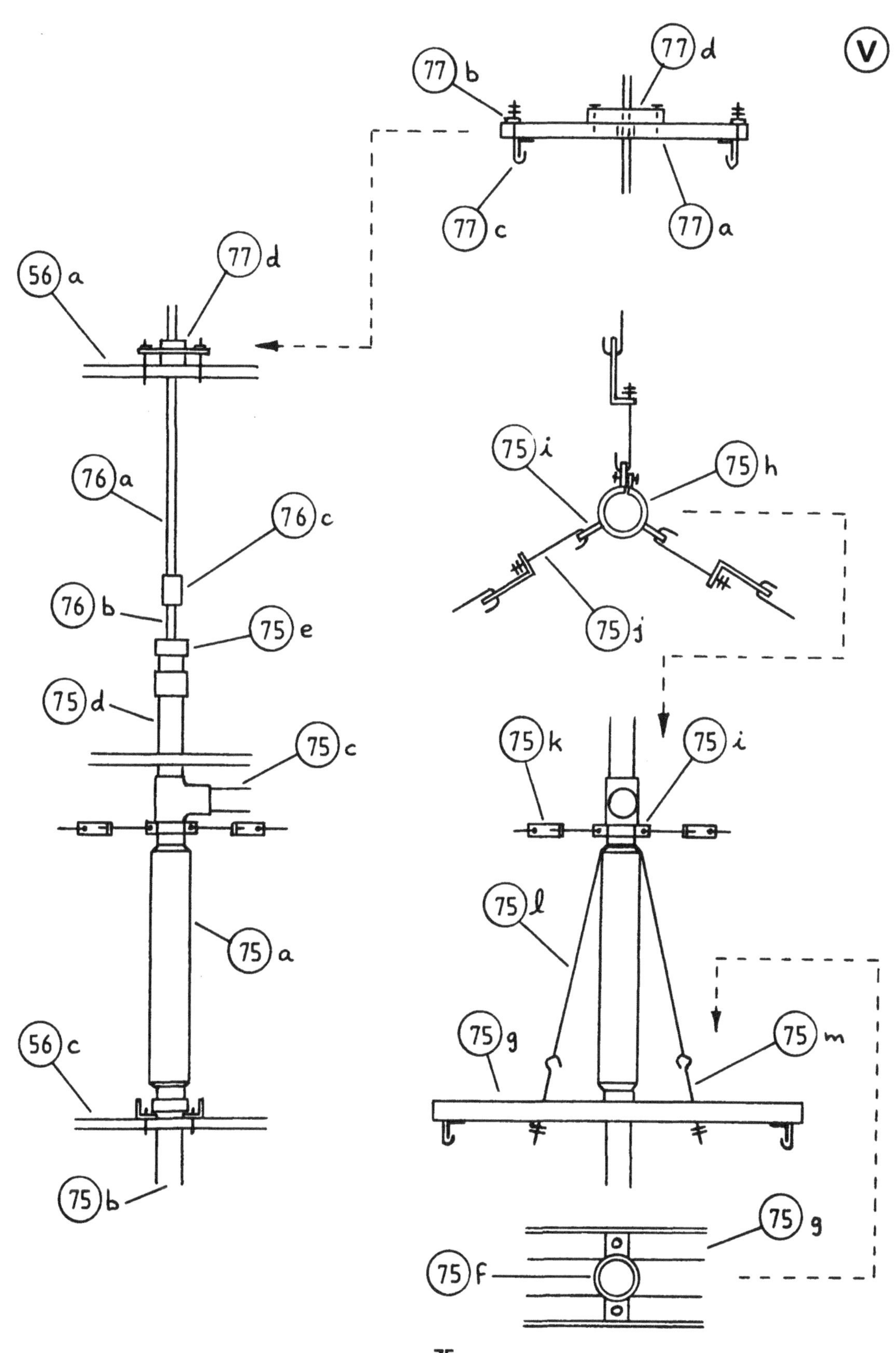
V
77 b
77 d
77 c
77 a
56 a
77 d
76 a
76 c
76 b
75 e
75 d
75 c
75 a
56 c
75 b
75 i
75 h
75 j
75 k
75 i
75 l
75 g
75 m
75 g
75 f

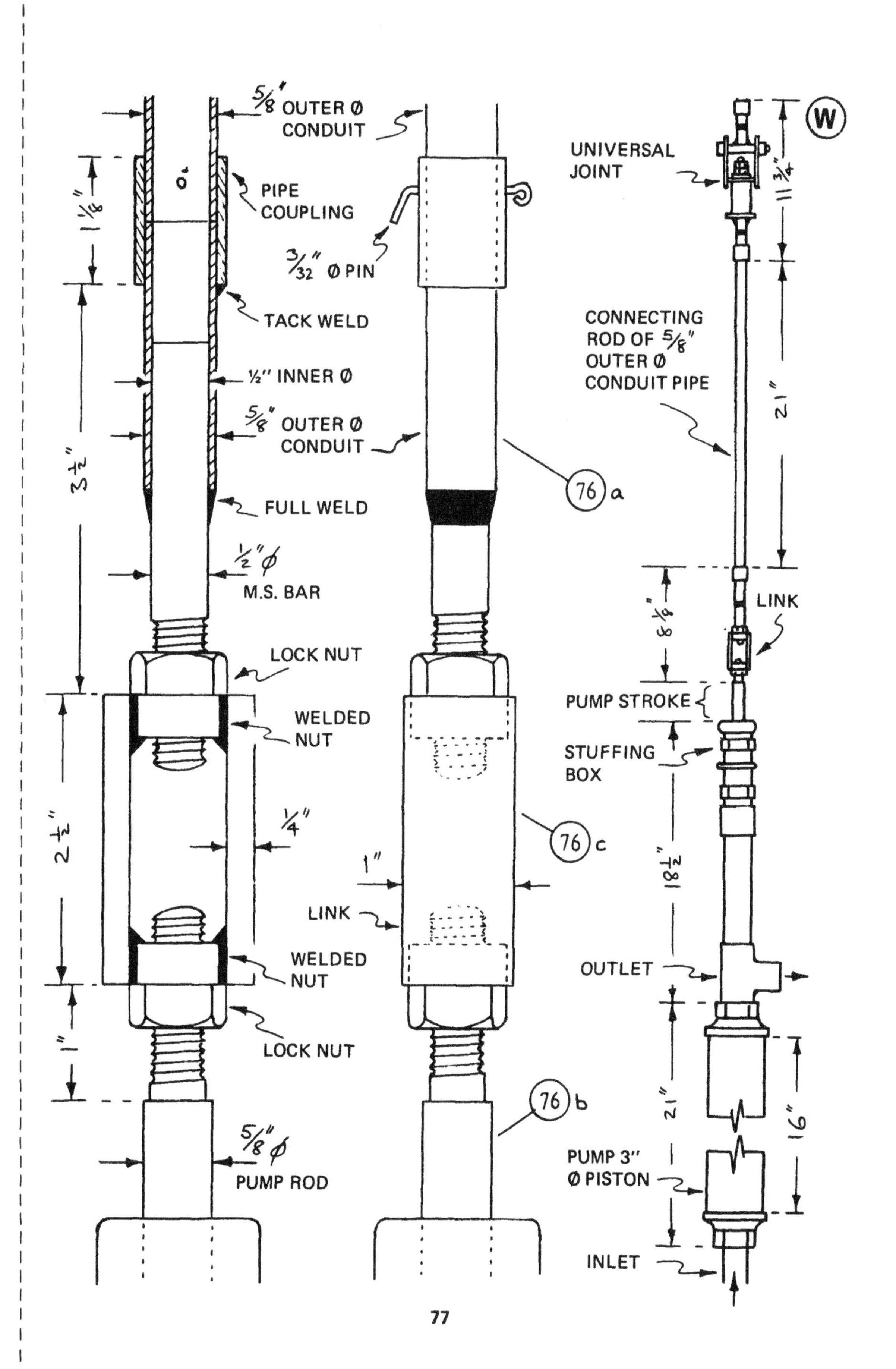
5/8" OUTER Ø
CONDUIT
PIPE
COUPLING
3/32" Ø PIN
TACK WELD
½" INNER Ø
5/8" OUTER Ø
CONDUIT
FULL WELD
½" Ø
M.S. BAR
LOCK NUT
WELDED
NUT
1/4"
WELDED
NUT
LOCK NUT
5/8" Ø
PUMP ROD
1"
LINK
76 a
76 c
76 b
W
UNIVERSAL
JOINT
11 3/4"
CONNECTING
ROD OF 5/8"
OUTER Ø
CONDUIT PIPE
21"
8 1/8"
LINK
PUMP STROKE
STUFFING
BOX
18 ½"
OUTLET
21"
16"
PUMP 3"
Ø PISTON
INLET
1 1/8"
3 ½"
2 ½"
1"

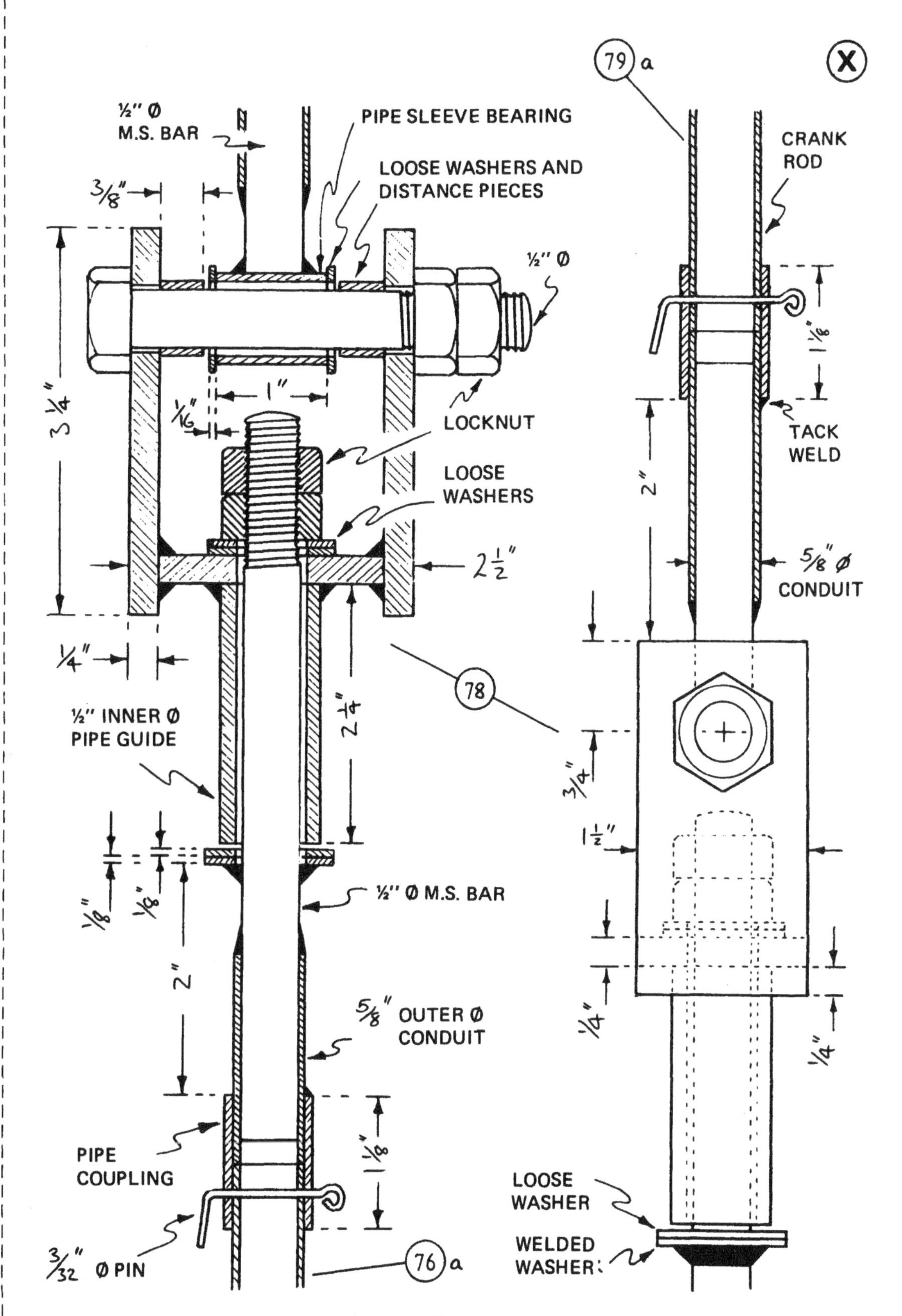

79 a
X
½" Ø M.S. BAR
PIPE SLEEVE BEARING
LOOSE WASHERS AND DISTANCE PIECES
CRANK ROD
3/8"
½" Ø
3 ¼"
1"
1/16"
LOCKNUT
1⅛"
TACK WELD
LOOSE WASHERS
2"
2½"
5/8" Ø CONDUIT
¼"
78
½" INNER Ø PIPE GUIDE
2¼"
3/4"
1½"
1/8"
1/8"
½" Ø M.S. BAR
2"
5/8" OUTER Ø CONDUIT
¼"
¼"
PIPE COUPLING
1⅛"
LOOSE WASHER
WELDED WASHER
3/32" Ø PIN
76 a

www.ingramcontent.com/pod-product-compliance
Lightning Source LLC
Jackson TN
JSHW062202130125
77033JS00018B/604
9780903031660